D0899038

ENDING PLAGUE

ENDING PLAGUE

A SCHOLAR'S OBLIGATION IN AN AGE OF CORRUPTION

Dr. Francis W. Ruscetti,
Dr. Judy A. Mikovits, and
Kent Heckenlively, JD

Skyhorse Publishing

Skyhorse Publishing books may be purchased in bulk at special discounts for
sales promotion, corporate gifts, fund-raising, or educational purposes. Special
editions can also be created to specifications. For details, contact the Special Sales
Department, Skyhorse Publishing, 307 West 36th Street, 11th Floor, New York, NY
10018 or info@skyhorsepublishing.com.

Skyhorse® and Skyhorse Publishing® are registered trademarks of Skyhorse
Publishing, Inc.®, a Delaware corporation.

Visit our website at www.skyhorsepublishing.com.

10 9 8 7 6 5 4 3 2

Library of Congress Cataloging-in-Publication Data is available on file.

Print ISBN: 978-1-5107-6468-2
Ebook ISBN: 978-1-5107-6471-2

Printed in the United States of America

It's the action, not the fruit of the action, that's important.
You have to do the right thing. It may not be in your power,
may not be in your time, that there'll be any fruit.

But that doesn't mean you stop doing the right thing. You
may never know what results come from your action.
But if you do nothing, there will be no result.

—Mahatma Gandhi

Contents

PART ONE

Frank's Perspective

Why Now?

The real safeguard of democracy is education.
—Franklin Roosevelt

Dealing with those vulgar souls whose narrow optics can see little but the little circle of their own selfish concerns.
—Robert Morris to Alexander Hamilton

The forced ending of my scientific career in 2013 was both personally and professionally disturbing to me.

However, it allowed me to join my wife Sandy, who was ready to retire after an excellent career in science, in our favorite place in the world, Carlsbad, California, a magical location next to the Pacific Ocean, just north of San Diego. To my pleasant surprise, the move also proved to be liberating. Most people reach a point in their careers where all the institutional politics and backstabbing hinder the creativity which first drew them to the profession. My absence from the National Cancer Institute (partly located on the grounds of the former United States Biological Warfare Weapons Laboratories at Fort Detrick) in Frederick, Maryland, my home for thirty-eight years, allowed me to reevaluate all the events in my career that rushed by in a blur.

I've grown to appreciate the truth of Allen Saunders's statement that "Life is what happens to us while we are making other plans." My understanding of what has happened to medical research in its application to public health in the overall context of American history during my lifetime has become dramatically clearer.

My career choice was to join what I considered to be an ancient and honorable society of scholars, which I joined in May 1972, upon earning my PhD. In this contemporary climate of increasing contempt for intellectual honesty, along with the delegitimizing of expertise, one may reasonably ask, why bother?

I believe we should "bother" because, as Gandhi's statement at the opening of this book said, the single most important obligation of a scholar is the production of knowledge. Knowledge in most fields, but most notably in science, has a long incubation period and has to be laboriously developed. Then, in a more difficult exercise, it must be communicated to a rightfully skeptical conservative audience, bound to the status quo. Skepticism is one thing, but I have found acceptance of paradigm-changing work by many medical researchers, more interested in protecting their own place in the hierarchy than in advancing knowledge, typically goes through a three-step process.

The first step is "no, you're wrong."

The second is "no, you're dead wrong."

The third is "I knew it all the time." This acceptance can take decades.

One of the more disturbing modern trends in science is the new cottage industry of completely twisting the truth for one's political agenda. Many of the results in scientific papers cannot be reproduced in the short term, mostly because of technical differences between the labs. The use of these facts by politically motivated citizens and scientists alike to deny science they do not like is often misused to discredit paradigm-changing science. This behavior is not only intellectually dishonest, but displays a complete misunderstanding of the scientific process. The rush to discredit these publications and even force retractions does a foolhardy disservice to scientific scholarship. In paraphrasing scientists from Darwin to Planck, a scientist should not fret over convincing one's peers. But instead, make certain the work appears in the next generation's textbooks. New knowledge that stands the test of time makes life sweeter in the succeeding generations.

The misuse of the scientific process by these individuals has the power to corrupt and cheat many brilliant and honest scientists of their rightful place in history. From the very beginnings of our history, Robert Morris and Alexander Hamilton, whose economic brilliance saved the American cause in the revolution and the new country, knew shallow, moneyed self-interests were the biggest threat to the republic. To whitewash their crimes and self-aggrandize their own personal achievements, the powerful elite have the ability to impugn and expunge the work of my collaborators, especially Judy Mikovits. While I am content in the knowledge that, we have made life sweeter for people, regardless of what my peers and their enablers may say, I am not comfortable that the mendacities and misdeeds of these unethical contemporaries go unrecognized.

Given the ability of objective facts to be twisted and turned into untruths, it's almost certain this will happen to most of what I say here, including my right to be called a scholar. The hardest thing to do is to know the value of one's own achievements, regardless of the opinions of others. Success in this world is often a mirage, the result of being praised by others or lavished with awards and money, regardless of whether the work has merit or not. Strive for achievement, not the praise of the world.

While struggling to develop a science career in the 1970s, it did not dawn on me that during the next fifty years an increasingly corrupt corporate apparatus was placing most people into economic slavery, where the important decisions concerning every aspect of our lives would be made by the rich elite, who are protected from any political or social consequences.

How did this happen? Corporate America is killing democracy.

The lion's share of the fault lies with the government whose duty it is to protect its citizens and instead allowed the development of crony capitalism, which is based on a close relationship between rich businessmen and the state. Instead of success being determined by a free market, it is determined by state favoritism in terms of tax breaks, little regulation, and grants.

Think about how different our world is now from 1970. Every aspect of our lives is controlled by the monopolization of corporate America, which makes it easier for foul people to control and undermine our freedoms. Banks are too big to fail, which is socialism for the rich. This has allowed corruption on a worldwide scale. This has led to our increasingly concentrated and corrupt medical system, which is literally killing us, led by the Food and Drug Administration (FDA), Centers for Disease Control and Prevention (CDC), Environmental Protection Agency (EPA), National Institute of Allergy and Infectious Diseases (NIAID), and the Department of Health and Human Services (HHS). Since the 1870s, the Republican Party has been a pro-business organization that corrupted the public trough and has given us J. P. Morgan, the monopolist banker, then Andrew Mellon, the robber baron/treasurer who caused the Great Depression, and Michael Milken, the "greed is good" junk bond king

The Democratic Party has joined them in becoming the world's second greatest pro-business party, completely disowning the working man, the middle class, and social justice. The strength of a representative form of government is that we the citizens can fix these massive problems. Most issues are so complex that considerable education is required to make decisions.

But do we have the guts to accomplish returning the government to the people? Do we have the guts to end the rampant corruption?

CHAPTER ONE

Science Saves My Life

An unhappy childhood compels you to use your imagination to create a world in which you can be happy. Use your old grief, that's the gift you're given.

—Sue Grafton

As a young boy, I knew nothing of the dark side of organizations that regulate an individual's life, liberty, and pursuit of happiness. I did not realize how much I would find myself in conflict with them over the course of my career. I love to collaborate with people of integrity, and nothing thrills me as much as a provocative question, the answer to which holds the promise of making the lives of millions of people better. But the organizations which cherish such values are few, and I worry that they are continuing to diminish.

My story is as ordinary in its details as millions of my countrymen. Charles Wildberger, my maternal grandfather, was born in New Jersey from parents who emigrated from Switzerland. He married Emma Steffe, whose parents immigrated to the United States from Prussia to escape conscription in the Franco-Prussian war of 1871. They had three sons and six daughters, including my mother Dorothy.

After serving in the Italian army as an ambulance driver on the Austrian front in World War I, my paternal grandfather Dominico immigrated to the United States alone in 1920. He worked as a machine worker to support his wife Cecelia, his daughter Clementine, age ten, and his son Frank, age fourteen, which eventually allowed them to leave Italy in 1928 and join him in America.

It was fortuitous that Dominico immigrated alone in 1920 because the Immigration Act of 1924 barred Southern and Eastern Europeans. Italian immigration dropped 90–95 percent. If he'd waited past 1924, I likely would not have been born, for my father and mother would have been separated by an ocean. Some of my eventual critics might have considered that a blessing.

Dominico always used the Ruscitti spelling of the last name. The first time Ruscetti was used was on my parents' marriage license in 1938.

Years later, my mother told me that Dominico had displayed the medals he was awarded for courage as an ambulance driver on the Austrian front during WWI on the mantelpiece. But as the Italian fascists entered World War II, he took them down and would never show them to me when I asked. He would only volunteer that war was bad and shake his head, a veil of silence descending around him. To this day, reading *A Farewell to Arms* makes me think of my grandfather. Both my grandfather and father, much to my regret, refused to teach me Italian. They both had seen too many "Italians need not apply" signs and said that, to get ahead in America, I should only speak English.

When I hear people speak disparagingly of immigrants today, I can only remember my family history. America was certainly better than where my parents had come from, but it was still far from being the "shining city on a hill." We so often think that people of all ethnic backgrounds have gained equality in America, but we're but a few generations removed from rampant racism, lynchings, physical abuse, and other types of injustice, which could return. Clearly, we have similar injustices today, just primarily aimed at different groups of people. Also, modern problems like the wholesale censoring of those who question the medical mainstream narrative have arisen. The veneer of civilization is thin, a fact we would do well to remember in 2021.

I was born on February 6, 1943, while the Russians were winning the European front of World War II in the streets of Stalingrad. Blizzards were also belting Boston, causing me to spend the first four days of my life in the hospital. Coming of age in Boston, a city rigidly segregated along ethnic lines with obvious class tensions, was both good and bad, with the Cabots, Lodges, and Lowells struggling to retain their power against rising Irish upstarts like the Curleys and the Kennedys.

Despite never having much money, there was plenty to do via inexpensive public transportation to Revere Beach, Fenway Park double headers for twenty-five cents, and NBA double-headers in the 1960s where one might see Wilt Chamberlain, Oscar Palmer Robertson (the "Big O"), or legendary Celtics point guard, Bob Cousy. The children's museum, then in Jamaica Plain, was a great museum, which pioneered a "hands-on" approach, letting us handle artifacts from different lands and investigate what most interested us.

Our home was on the second floor of a three-floor, six apartment rental. My parents could never pull the trigger on a home purchase and, early on, we did not even have a car. One of the many myths we are served as children is the myth of the happy childhood. Parenting is difficult and a lot must be given up in order to raise children.

Regrettably, I never attempted to understand my parents' perspective while I lived under their roof.

My mother was perpetually unhappy. Raised voices and constant arguments were the background noise of our lives. In such a small apartment, it was nearly impossible to get any peace and quiet. We three kids never got the space we needed as we got older. I was not a particularly brave child. I was afraid of dogs (being bitten three times and needing a rabies shot one of those times) and terrified of fire (a neighbor's child burned to death in a Christmas tree fire. I can vividly recall the child's screams to this day.)

Whenever my dad tried to teach me to ride a bike, swim, or drive a car, my mother would scream at him that I would get hurt and he'd give up. I still wonder what events in her life seemed to make her so fearful of the world. My father and I shared a room. Since he worked a lot of double shifts on the New York, New Haven, and Hartford railroad freight lines, I spent a lot of my spare time alone in the room. I'd get lost in listening to the newest musical trend, rock and roll, as well as the broadcasts of the Red Sox and Celtics games. On Sundays, I listened to the broadcasts of WHDH, which carried the Cleveland Browns football games featuring players like Otto Graham and Jim Brown.

I started school at four and a half years old, and it offered me a chance to get out of the house. At Nathaniel Bowditch Elementary School, I encountered my first conflict with institutional authority. From the first to the fourth grade, the teachers tried to get me to write with my right hand, instead of my left.

I refused.

As punishment, I was made to sit in the last seat in the last row and it was my job to fill the inkwells. From the very beginning, I resolved not to obey insufferably mindless authority. At a parents' classroom meeting with my first-grade teacher, she mentioned that she would assign the smartest student the duty of passing out the napkins. I immediately began passing out the napkins, even though the task had not been assigned to me. My mother, of course, was mortified.

As much as my mother was terrified of the world, she had no trouble embarrassing me and putting me in actual physical danger by forcing me to wear "I like Ike" ties to school in democratic Boston. Every year on March 17, Saint Patrick's Day, she made me wear something red, white, and blue for "Evacuation Day," commemorating Washington driving the British out of Boston in 1776, rather than celebrating the Irish. I took lots of verbal abuse for that, too.

Would it have killed my mother to let me wear something green on Saint Patrick's Day?

Several incidents during my childhood affected my personal outlook. My sister Barbara developed glomerulonephritis (an acute inflammation of the tiny filters of the kidney) as a side effect of a sulfa drug prescribed to treat bacterial infections and had to spend a prolonged period of time in bed. In 1953–1954, there were local polio outbreaks and our mother would not let us go outside at all that summer. The pictures of children in iron lungs were enough to scare anyone. When the Salk vaccine became available, my mother asked me to take Barbara and Bob to our family physician, Dr. Beale.

He refused to give it to us, saying it was useless.

My mother, severely irritated with me, sent us back and we got the vaccine. Shortly thereafter, one of Dr. Beale's children came down with polio. Thus, I learned medicine was not exact, and doctors were not infallible. This would prove to be one of the recurring themes in my life.

One night, police pounded on our door, dragged my father and me (all of eleven or twelve years old) out of bed, put us up against the wall, and frisked us. They were looking for a drug dealer named Frank Russo. They left without saying anything like "sorry," further degrading my opinion of the police. Around the same time, a parish priest came to the door and told me to tell my parents that since they hadn't been married in the church (they'd been married in a civil ceremony), that they were living in sin.

I promptly told him where to go.

By an early age I'd already developed a skeptical attitude toward authority figures such as teachers, doctors, police, and priests.

And yet, even with my attitude, when the organizations were well-run and rational, I could thrive in them. My time at Mary E. Curley Junior High in Boston (grades seven through nine), were among the best years of my life, until graduate school. It was a source of great pleasure and satisfaction when the students in my class said to the teacher, "Don't ask us the answer. Ask Ruscetti." Combined with gym class, basketball, public library study sessions, Saturday morning movies, Congregational Church Sunday school, and Sunday evenings in Christian Endeavor, a good refuge was formed for this young man.

I still remember the two teachers (Mrs. Fodale and Mr. Cannon) who wanted me to apply after eighth grade (a year early) for the entrance exam to Boston Latin High School, because they thought I was so advanced. However, it was a long streetcar ride away and my mother refused, saying I was too young to take the trolley. Later, I took a similar entrance exam at the end of ninth grade and was admitted to Boston English, ironically riding the same trolley, only a year older. Meanwhile, at our apartment building I oversaw planting, pruning hedges, and shoveling snow, resulting in a monthly reduction in our rent.

Years later, when rock and roll artists Bob Dylan, John Lennon, and Paul Simon sang about what a waste of time high school was for them, I couldn't have agreed more. Homework had to be done during school study periods, because neighborhood bullies would target you for taking too many books home. My favorite high school teacher, a math teacher who for some reason had been dismissed from teaching at the US Naval Academy, taught a fascinating course in navigation and meteorology. Sadly, another favorite, Mr. Ruggiero, died young of leukemia.

I was fourteen years old when the Russians launched the first satellite, *Sputnik*, on October 4, 1957. America changed overnight as we were terrified that the Soviets were going to beat us in the space race. The US government got the education it wanted. All students had to participate in military drills at my high school and know where the fallout shelters were located. We believed the evil Russians were going to destroy our perfect country.

The worst of my high school experiences was the complete absence of any useful guidance counseling when it came to college applications. With little money available, I assumed that living at home was the only option. So, I originally applied to just Boston University and that other university across the Charles River. A representative of that university told me that the Italian quota was filled. Boston College was not an option for me. Anyone who has seen the Oscar-winning film *Spotlight* knows the school's toxic influence on Catholic Boston.

One of the best things that came out of the 1960s social unrest was that the elite WASP schools were forced to open their enrollment to most everyone. Another thing which greatly irritated me was later learning there were scholarships donated by wealthy high school alumnae, available for Dartmouth and Bowdoin, for which I would have applied, if I'd been told. How could I compete if I didn't even know about the opportunity?

Boston University was both a cultural shock to me and a disappointment. As a result of being shy and from an all-boys high school, an aunt had started a family rumor I was gay. I mistakenly registered for a freshman composition class in the School of Nursing. I was the only male in the class, leading to plenty of blushing. For the qualitative analysis lab final, in which we were supposed to determine the chemicals inside the test tubes, I was mistakenly given test tubes with the answers already written on them. I turned it back in for a new set of test tubes.

One science professor stated that he gave an A to God, B to the smartest student in the class, and a C to everybody else. When I complained, he said I should be happy with my B-. The organic chemistry professor graded on a downward curve, so my 95 average became a B+.

So much for the lunacy of academic grading.

Petty dictators were everywhere you looked.

There were some great college memories as well. I watched Gale Sayers of Kansas score his first collegiate touchdown to beat BU, 7-0, attended my first American Football League (AFL) game, Denver at Boston, and I saw Faye Dunaway, the future Oscar-winning actress, at a BU theater event.

Most of what I learned came from my summer jobs. Through my dad, I got a job at the South Station railroad station slinging hash. I learned serving customers was not going to be my forte. Then I worked for the state of Massachusetts on a crew repainting crosswalks and center lines. Occasionally they'd go to a house of ill repute on their lunch hour, where I had to wait outside because I was underage. I didn't fit in there, either.

My uncle, Bill Wildberger, was a family hero. He was the first non-Harvard graduate to become chief resident at Newton-Wellesley Hospital. Later, he was the mental health director for the state of Iowa and got me a job during college at the state mental hospital in Woodward, Iowa. There was a hepatitis outbreak the first week on the job, meaning I had to perform X-rays, draw and run routine blood tests, run the pharmacy, and assist the pathologist when she arrived from the state medical school. It was hectic and nerve-wracking.

But for the first time I thought I could see how I could make a difference.

It was appalling to see how inborn genetic errors could cause such pain in people. I sympathized deeply with these people, although they often terrified me. Many times, security had to come to my rescue because the patients would try to assault me while I was attempting to draw their blood.

Iowa was a dry state and Woodward had no movie theaters or even bowling alleys. Once I asked some of my female coworkers what there was to do for fun in the area. They answered, "watching the corn grow!" Apparently, a shy city boy needed an education.

Iowa didn't seem like the place for me, either.

My final job during college was repairing railroad tracks in the blistering summer sun. It was backbreaking work. I was the one summer hire who made it through the entire summer, a source of immense pride to me. (My father had gotten me the job, so I couldn't let him down.) After the summer, my dad said in his laconic fashion, "There are two ways to make a living. With your back or your brain."

I've often wondered if I made the right choice.

The working men usually seemed more honest than most of the professionals I've encountered in my career. However, my father also told me that several times he had wanted an opportunity to take a test to become a freight yardmaster, but he was refused the opportunity because it was assumed an immigrant could not pass it. Finally, he was allowed to take it and passed, receiving several commendations for his yardmaster work over the years. He warned me to do my work better than anyone else: "Let them find something to complain about, but not the work!" Like many sons, I have tried to emulate my father's virtues and avoid his flaws.

Through the years, I have talked to many people from poorer backgrounds in urban settings who had to commute to college, as I did. I recognize now that it was not commuting per se, but the claustrophobic home environment of so many that was the problem. Despite being belatedly accepted to the University of Virginia Medical School after being waitlisted, I decided I needed a change. I have always admired my brother and sister, who made well-adjusted lives out of such chaos.

My solution?

Join the Air Force.

Probably not a wise choice for a young man who in first grade vowed not to obey mindless authority.

The military was quite a learning experience. Two lessons which stand out are: First, the dangers of small men in positions of power; and second, war is the most unfair and idiotic of the many foul endeavors in which man participates. Basic training was barely tolerable with the constant screaming of the drill instructor.

It reminded me of home.

Half of my basic training group had Boston accents and the other half were North Carolina tobacco boys. We could barely understand each other. A good percentage of our flight squadron was ordered to Montgomery, Alabama to attend medic school.

None of us could remember stating that as a choice.

Montgomery was not a good place for a Yankee like me to be in 1965. Every store owner had a rocking chair and a rifle. Southern boys usually had enough munitions in their car trunk to conquer Mexico. Every time we went out to eat, the southern farm boys ordered for me so I wouldn't get shot. One day in class, the instructor asked if anyone knew how to run a Model E Ultra-Centrifuge.

No hand went up, so I raised mine, thinking that I'd used centrifuges before. How could this one be much different?

But this one looked unlike any I'd ever seen and took up half a wall. Thanks to blind luck and pushing the right buttons, the centrifuge performed smoothly. Next thing I knew, my personal folder was stamped "ESSENTIAL TO SPACE PROGRAM."

New orders shipped me to Lackland Air Force Base, the main US air evacuation hospital for injured soldiers in San Antonio.

Early in the space program, it was discovered that the red blood cells of the astronauts lasted only ten days instead of the normal twenty-one. They wondered if that was going to be a long-term problem, thus complicating any planned trip to the moon. The answer turned out to be no. In a few weeks, the red blood cells recovered their normal life span.

But it started my long career fascination with hematology, thanks to instruction from Dr. Chuck Coltman, Dorothy Grisham, and others. Knowing some friends who did not make it back from Vietnam, I've often wondered whether science saved my life.

I needed to get up at five-thirty in the morning to draw blood from the injured soldiers and had to finish before going to the mess hall for breakfast. I missed many a breakfast before officially reporting to work at seven-thirty. Some would forget to do their blood requisitions, discard them, and others were not that good at the task. Of course, those of us who completed our assignments started to get a bigger portion of the workload. Drawing blood from napalm victims (our own troops often had napalm bombs accidentally dropped on them) on the burn wards was the absolute worst.

After lab classes, the remaining time was supposed to be devoted to research endeavors. But the major in charge kept finding more and more for me to do, like preparing and changing solutions for dialysis patients.

Again, there can be so many petty dictators in life.

On weekends, I would be part of a team (which I later supervised), which would draw four hundred units of blood to be sent to Vietnam. Habitually tired, I'd often fall asleep in hematology lab class. One day I was asked if I thought I could teach the class better and got in trouble because I could not lie and thus said "Yes" as I walked up to the front of the class and was rudely sent back to my seat.

The major in charge also thought my hair was too long, so he could often be found prowling around the lab, surprising me to check my hair length. Apparently, the length of my hair was critical to the success of our war effort in Southeast Asia. To humiliate me further, the major would often have military police escort me to the base barber shop, so that as many people as possible would see me walking through the halls under military escort. It was just like being back in elementary school where the teachers would inflexibly try to make me stop writing with my left hand or like that night being thrown up against the wall because the police thought my dad was a drug dealer.

You may not believe it, but I did spend most of my time in the service trying not to get in trouble. Wounded servicemen were always increasing in numbers. General William Westmoreland (in charge of the US effort in Vietnam) was always saying at commander call (a required meeting for all enlisted personnel) that the casualties were going down. He'd talked about there being "light at the end of the tunnel," but we always joked that the light was a train coming to run us over. I consider General Westmoreland to be one of the biggest liars in American history.

And sometimes it seemed the hypocrisy knew no limits.

On the parade grounds, we tried not to smile at a Purple Heart ceremony where the soldiers had been injured when the Viet Cong blew up a whorehouse where the troops had been engaged in a little "rest and relaxation."

Or being ordered to spend everything left in the budget days before the fiscal year ended so we would not lose it in next year's budget.

I tried not to be dismayed when a black enlisted man and a white lieutenant nurse went to a movie with me on base, and then the next day he had orders to go to Vietnam, and she was sent to Thule Air Base in Greenland.

When the *MASH* television show came out in the 1970s, depicting the absurdities of life at a military field hospital near the front lines of the

Korean War, I thought, *I've been there*. It's no surprise it was one of the most popular television shows of the decade, and its final episode was watched by a reported 125 million viewers.

After it became clear that the lower red blood cell survival rate of our returning astronauts was just a transient problem, and the red blood cells would eventually rebound to their normal twenty-one-day survival rate, I received orders to report to Oxnard, California.

However, my orders were mysteriously cancelled.

The only option was to plan my escape. But what to do?

Science or medicine?

They are not the same thing.

My choice was science. I reasoned that if I were lucky enough to discover something useful, it would help people long after my death. At the time, a National Institutes of Health (NIH) graduate school stipend was $2400, while an Andrew Mellon fellowship, usable at any school department that had an Andrew Mellon professor, was $3,200 a year. An extra $800 dollars a year was a lot of money to someone who made only $1,300 in almost four years of service. And there was an opening at the University of Pittsburgh.

So I took the money, which was the legacy of Andrew Mellon, the former oil and steel tycoon, who I'd later come to realize was one of the most evil and powerful men to ever live in America, and headed to the University of Pittsburgh. I flew standby in my uniform, not having any money on me, because I needed to physically appear at the university to get my first check. I remember explaining all of this to the bell hop at the Webster Hall Hotel when I checked in, telling him I was broke and could not give him a tip.

He grinned and said, "We've all been there, son," and closed the door to my room. I never gave any thought to what lay ahead of me; I was happy to escape the senseless suffering of war.

The allure of science, of discovery, of doing something useful, was tantalizing.

CHAPTER TWO

Protest and Pittsburgh

I have almost reached the regrettable conclusion that the Negro's main stumbling block toward freedom is not the Ku Klux Klanner, but the white moderate who prefers 'order' to justice.

—Dr. Martin Luther King Jr.

The times they are a changin'.

—"The Times, They Are a-Changin'," Bob Dylan (1964)

The day I arrived in Pittsburgh at the end of August 1968, having dropped my bags off at the hotel, I rushed to Langley Hall (the arts and sciences building of the university). I didn't even bother to change out of my uniform.

The halls were mostly empty, but I was able to find the departmental office and pick up my check.

By chance I ran into Dr. Lew Jacobson, a young, newly minted assistant professor. He would become my PhD advisor, as well as a lifelong friend. We must have talked for an hour that day. I told him about my experiences in the Air Force, we discussed the ongoing riots at the Democratic Party Convention in Chicago, pick-up basketball, and graduate school life. As far as where I might want to live, he suggested the Shadyside district, just about fifteen minutes from campus, and I quickly found an apartment in that neighborhood.

The first year of graduate school was a rush of academics and performing my duties as a teaching assistant in an undergraduate biology class. In my naïve idealism, when a female student said, "I'd do anything for an A," I replied, "Try studying."

When Richard Nixon was elected over Hubert Humphrey in November 1968, by appealing though a southern strategy to white supremacists and the law and order, pro-war crowd, I was crushed. I had truly believed as Bob Dylan sang, that the times were a-changing.

Years later, when it was shown that Nixon had sabotaged the Vietnamese peace talks to help win the election, along with Watergate and Nixon's secret wars in Cambodia and Laos coming to light, I wasn't the least bit surprised.

It set a pattern for imperial presidential misconduct through my lifetime. This is not a political statement. I was just observing a social progression toward an autocracy which bedevils us today. When Bob Dylan released a song in 2020 called "Murder Most Foul," talking about the corruption that followed for the next five decades in our political system after the assassination of President Kennedy in 1963, I thought the old singer was still in touch with the times.

As fate would have it, several personal blessings came to me during these troubled times in the life of our nation. On February 2, 1969, at a graduate student party, I met Sandra Ickes, another aspiring scientist.

Meeting her felt like finding the missing part of my soul.

One does not get to my age without having a clear picture of his faults. I know I can be moody, uncommunicative, and judgmental. Yet the part many do not see is that it's because I can imagine a much better world. Much of the time I have felt like a disappointed idealist. Others have described me with less kind names, and I don't know if I can blame them.

But Sandra did not have my darkness.

She was social, saw the best in people, and had no trouble making friends. She was like oxygen and I was the drowning man she could save. Imagine how thrilled I was when, four days after that party, I received a birthday card from her. I had simply mentioned my birthday was in a few days in passing at the party. When I thanked her, believing it made me something truly special, she almost crushed my fragile heart when she replied, "Oh, I send them to all my friends." But I was hooked.

I invited her to another party, but she turned me down. Later, I learned she'd been asked by another student, had turned him down, and didn't want to make the other guy feel bad by showing up with me. But I persisted, and soon we were a couple.

Even after all these years, my lovely Sandy has a remarkably forgiving soul, the optimism of her Midwestern upbringing, and an honesty which is second to none. When Sandy and I traveled to Boston to meet my parents, my family's admiration for her was in full force. My mother said she did not know me. I could have been a bank robber. My father told Sandy that she could do much better than me. Great family support!

We married in April 1970 as the Beatles song "Let it Be" played on the organ. I can still recite some of the lyrics, with the refrain stamped forever on my soul ("there will be an answer, Mother Mary"). We had our honeymoon on the Caribbean island of Aruba and returned to the University of Pittsburgh full of hope for the future.

Another fortuitous event was the selection of my advisor for my doctoral research, Dr. Lew Jacobson, whom I'd met on my first day at the University of Pittsburgh. I say you should always try to choose a mentor who is superior to you in every way, which I did.

Well, maybe I was a better basketball player than Lew.

But that's about the only superiority I can claim over that fine man. I chose to study a project using bacteria, which was great because under proper conditions the microorganisms doubled every twenty minutes, allowing me to perform at least two experiments in a single day. That fit nicely with my plan to finish my degree in four years.

Every morning I'd enter the data from the results of the experiment from the previous night in my notebook and looked forward to several hours of analyzing the results. However, each day, Lew would have written out enough experiments for the next two weeks' work. So I chose to take my notebooks home to ponder the results and plan the next day's experiments. It was always a treat when I found my approach matched that of my mentor.

I've often found that failure is a better teacher than success and recall the shock many of us felt when our graduate class failed our first preliminary exam. Did they plan it that way just to show us how much we did not know and keep us humble?

Even fifty years later I don't know the answer to that question.

Perhaps the major downside of being a scientist is the constant struggle for funds to do research. Learning to manage resources was a valuable lesson. But there were the positive lessons as well, such as the camaraderie between colleagues at the weekly Friday night graduate students' pizza and beer parties at the Craig Street Inn, where we'd get together and bitch or laugh over what had happened the previous week. We wanted to be scientists after all, trying to answer the unknown questions of life. We needed to have each other's backs.

Most mornings I'd walk past Central Catholic High School between 5:30 and 6:00 a.m. on my way to Langley Hall. The early morning start times were a legacy from the 5:30 a.m. blood draws in the Air Force and have followed me all the days of my life. I'd often see a young boy at the school, usually with somebody else he'd no doubt dragged out with him, throwing a football around. That young man was Dan Marino, future NFL quarterback and one of the best to ever play the game.

Sandy and I often rode the same bus as Franco Harris, an NFL rookie, as he was on his way to practice at Pitt Stadium. I wonder how many NFL rookies use public transportation today? Harris is best known for the "Immaculate Reception," which took place on December 23, 1972 during the AFC divisional playoff game between the Pittsburgh Steelers and the Oakland Raiders. In the last thirty seconds of the game, Steeler's quarterback Terry Bradshaw threw a pass to receiver John Fuqua. The ball either bounced off the hands of Fuqua or the helmet of Raiders safety Jack Tatum. As the ball fell, Harris caught it (Did it touch the ground or not? Football purists still argue over that question) and ran it in for a touchdown.

That play has been chosen by the NFL as both the greatest, and also the most controversial, of all time.

I think it's a perfect symbol for the randomness of human experience.

* * *

Every science graduate student must pass a nerve-wracking comprehensive oral exam. And if that's not enough stress, mine was interrupted by a bomb scare and we had to evacuate the building.

The extra time didn't go to waste though, as I took the opportunity to formulate a more precise answer to the question which had just been asked of me. The political turmoil and violence of that time is difficult to imagine, although the recent presidential election of 2020 might give some younger people an inkling of what we experienced then. I recall at one anti-war protest at Point State Park that the pro-war protestors broke through the mounted police barricade and tried to beat us with an American flagpole. In writing, such a scene might be called too "on the nose."

I can imagine some editor somewhere saying, "No, Frank, the symbolism of peace protestors being beaten with an American flag is just too obvious. Come up with something a little subtler." But that's how I remember it. The image has never left me and flooded my memory this past January 2021.

Sadly, the times are not changing.

As the old proverb goes, "When the winds of change blow, some build walls and some build windmills." In our history, the will to change dissipates quickly and we go back to our comfortable lives. We could hope this cycle will not repeat itself in 2021, but the only way to realize this hope is to be that change. That is why we write this book.

In May 1972, I earned my doctorate and take pride in the fact that the resulting publication is still occasionally referenced in recent manuscripts. Normally one would think this would be a cause for celebration in a family. But my mother, true to form, proclaimed that I became "the wrong kind of doctor" just to spite her. Sandy had just changed research advisors, and as a result it would take her another two years to finish her doctorate. This is not unusual. I've found a mentor-student relationship can often be more complex than a marriage.

While Sandy was still busy working on her doctorate, it was time for me to find a job. I heard Dr. Dane Boggs, head of the hematology division at the University of Pittsburgh Medical School, give an exciting lecture about their work. He said the field of blood research needed more PhDs. I applied for and was hired as a research instructor for the whopping sum of ten thousand dollars a year, which was a three-fold increase over the stipend I'd been living on as a graduate student.

Ever since my days in NASA and that centrifuge, I have been fascinated with the study of blood. The ability of hematologic (blood) stem cells to produce large quantities of functional cells of at least twelve different cell types was a source of enormous fascination for me. In science, technological advances generally precede intellectual advances in science. In essence, we are almost always blind to reality until we are given new eyes with which to see.

In hematology, a key technological advance was the 1966 publications by Ray Bradley in Australia and Leo Sachs in Israel which described the ability of blood stem cells to grow and develop into functional blood cells as clumps (colonies) in semi-solid cultures, if provided the appropriate mix of nutrients and chemicals, usually called "media" as a source of factors.[1] Years later I met Bradley and found him to be a generous and humble man, the very model to me of a good scientist. At the University of Pittsburgh, I chose to learn Bradley's technique from Dr. Paul Chervenick and his technician, Joan Turner (née Allulunis), who later went on to a successful career of her own in science.

However, as I started to investigate these questions of which growth factors were needed to grow certain cell types, I began to realize my illusion that science promoted a completely free discussion of research data and its implications was being slowly eroded.

There were unspoken rules of the game, and I didn't know them.

One of the unspoken rules was that the truth wasn't determined by your data, but by your seniority. An example of this is publishing data. It is extremely difficult to get data published and recognized when one is publishing second and correcting errors in the first publication. The first publication on a new topic became unassailable dogma, difficult to amend. Reviewers of future papers or NIH grant proposals would cite the first publication as truth and reject the second publication, especially if the investigators of the correct publication were young or not part of the club.

In the case of the identification of the factors needed to grow blood stem cells, the unassailable dogma was that the factors needed to grow these cells were produced only by cells called macrophages. Testing this hypothesis, we found a cell type called T lymphocytes produced such factors. It took nearly three years, several rejections and revisions, and many colleagues arrogantly challenging the data, before it was published.[2]

However, before publication, I made the mistake of presenting the data in both oral and written form at a prestigious clinical investigator meeting. Another investigator took pictures of the data, and his publication on the same subject beat mine into press by a month. After I presented my paradigm-changing data, which would open new areas of research, the first question I received was, "Why aren't you wearing a tie?"

I was learning that science could be a tough game. So, I resolved never to talk about data until it was in press and to buy a tie.

Another question we pursued was: Since there were so many different blood cell types, did each one of them need specific cell growth factors?

We addressed this question using cells grown in conditioned media from animals infected with parasites which developed eosinophilia. Doing this, we in the Chevernick lab were able to grow functional eosinophils in culture for the first time. Eosinophils are a type of blood cell which combats parasites and fungal infections, but also makes allergies and asthma worse. We named the factor the Eosinophil Growth and Differentiation Factor.[3]

My friend and future collaborator, Steve Bartelmez, confirmed the discovery using uninfected cells. Five years later, the name was changed from Eosinophil Growth and Differentiation Factor to IL-5, when another scientist cloned and sequenced the gene for the molecule. But in what was becoming an all-too-familiar pattern, that scientist was constantly disparaging the research Steve and I had conducted and claimed credit for the discovery of

IL-5. As a result of IL-5, I rewrote a failed grant of Dr. Chevernick, which included in it a position for me at a higher salary. I never knew the outcome of the grant review until another scientist showed up to take the position meant for me. Mike Kolitsky, the person hired for my position, turned out to be a great guy and long-time friend with a nice career in science education.

The mentorship and collegiality I experienced at the University of Pittsburgh was almost too good for the real world. Yet, the downside of science was starting to creep into my psyche. I was starting to believe that education did not change a person's basic character, just allowed for the more clever expression of it. Some scientists were going to be the most unethical people on the planet. Others would be the opposite.

Sports has always been an outlet which kept the corruption from overwhelming me. I found great relief in physical activity. Every day at lunch a bunch of guys would go to the gym and play an intense hour-long game of basketball to blow off steam.

The university was starting an intramural softball league for women and asked me to coach a team for the medical school. Few of the members, whether students, technicians, or professors, had ever played softball before. I encouraged Sandy to play and joked that it gave me an excuse to yell at her and she couldn't say anything about it. I had great fun coaching that team for two years.

Several of my team, who worked on the ninth floor in the surgery unit, introduced me to their boss, Dr. Bernie Fisher. To call Dr. Fisher a great man is an understatement. In the early 1970s, he published data suggesting that a lumpectomy was as good, if not better, for some breast cancer victims than the radical mastectomy, which was the "standard of care" at the time.[4] The savagery of the attacks on Dr. Fisher's character for suggesting less radical surgery was so stunning that one might have thought he had advocated for the cannibalism of young children. In the midst of facing multiple accusations of scientific misconduct, Fisher persevered, conducting several clinical trials that supported his theory. I remember him telling me and others that, "For far too long, medicine had depended on anecdotes, opinions, and untested theories." A favorite quote of his, which I later put up in my own lab at the National Cancer Institute, was, "In God we trust, everyone else better have data." I often wondered how many women were needlessly butchered by the refusal to accept paradigm-shifting science, even after their claims had been proven by clinical trials.

I'm embarrassed to say it never dawned on me in my early years that women in science were treated differently than men. Some of the best graduate students I'd met at that time were women. I figured we were all treated like crap, since we were the junior scientists. One day, I received a call from Dr. Sharon Johnson, one of my favorite graduate school instructors. She informed me that she'd filed a discrimination suit against the medical school biochemistry department for denying her a promotion. They claimed much of her work was inadequate, alleging she was a poor instructor and failed to publish in top journals. I agreed to testify for her. The night before the trial was to begin, I received a call from the dean of the medical school, warning that if I testified it could, "severely damage my career."

I did it anyway.

The university came off looking like a fool in the trial. One of the allegations from a fellow professor was that she didn't publish in the leading journals. Most of her publications were in the *Journal of Biological Chemistry*. The editor-in-chief of that journal was the chairman of Dr. Johnson's department. That meant the university was claiming the chairman of its own department was heading a poorly respected journal. Dr. Johnson won the trial, but as you can imagine, the environment was so toxic against her that she left the university. It was a great loss, as she was a gifted teacher and researcher.

From this and other experiences, I've concluded that corrupt patriarchal systems allow predatory men to believe they can get away with anything. In addition, I've come to believe that the more educated, wealthy, or successful a man is, the greater the chances are that he will be a sexual predator.

The recognition of how women were being treated poorly in science and that the unrest of the sixties was still doing little to resolve lingering racial problems was deeply unsettling to me. I remember talking about these problems with Georgia, the wonderful African-American dishwasher for the laboratory, when she said to me, "Frank, you're too impatient. It's always two steps forward, one step back."

* * *

While I was thoroughly enjoying my hematological research at the University of Pittsburgh, Sandy was finishing her doctoral work and I thought it was time for us to move on to more senior research positions. She was interested in working in the National Cancer Institute's intramural program and had attended a seminar given by Dr. Robert Gallo. She realized Dr. Gallo was interested in the same kind of research questions I'd been pursuing and suggested I put in an application with him.

In December 1974, Gallo and I were both attending a meeting of the American Society of Hematology, and he suggested we meet. I thought it would be a routine job interview, but instead it was more like a high-pressure sales pitch from Gallo.

"Don't you want to be working in a lab that will someday win a Nobel Prize?" he asked me at several points during our meeting. That alone should have been a red flag.

His other pitches to me were that I could study the growth factor that caused leukemic stem cells to grow continuously in culture, study the virus which caused acute myelogenous leukemia, and have a salary that was a considerable increase from my current one at the University of Pittsburgh.

We found an apartment in a nearby complex where several scientists at the National Institutes of Health (the parent institution of the National Cancer Institute) lived. In fact, so many of the researchers lived in this complex that there was a shuttle bus which ran between it and the various parts of the NIH campus. I occasionally found myself seated next to Dr. Julius Axelrod, who won the Nobel Prize in Medicine in 1970, and we had some interesting and lively discussions.

But while there were many scientists from whom I would learn exciting and revolutionary ideas, it was at the hands of Dr. Gallo that I was about to get an Ivy League education in unethical behavior.

CHAPTER THREE

National Cancer Institute:
Discovery and Disillusionment

*See how fortune deludes us and that which we carefully put in her
hands, she either breaks or causes it to be removed by the violence of
another, or suffocates and poisons, or taints with suspicion, fear and
jealousy to the great hurt of the possessor.*
— Giordano Bruno

I arrived at the National Cancer Institute (NCI) in March 1975 and was
stunned to discover that most of what Gallo had promised to me in
December turned out to be a lie. I later realized that when he spun out his
elaborate sales pitch of my future work, he knew it to be false. To understand how this happened, some background is necessary.

The confluence of the NCI's 1964 Special Viral Cancer Program,
(SVCP)[1] and Nixon's War on Cancer Act, passed in 1971, meant that more
money was available through contracts outside of the peer-review grant
mechanism funding process than ever before. The first year, the budget
was forty-nine million dollars for the NCI (an amount equivalent to about
$320 million in 2021), and the budget increased every year after that. This
caused a great deal of professional jealousy among scientists who weren't in
the government loop. The structures by which these financial decisions were
made tended to be concentrated in the hands of relatively few government
scientists/bureaucrats, like Gallo. For the first time, scientists were able to
build large laboratory empires using the public trough rather than going

through the peer-reviewed grant process of slowly building their labs as their colleagues (peers) agreed they were pursuing meaningful questions.

The Nixon administration was essentially building a scientific structure guaranteed to become corrupt because it concentrated absolute power in the hands of relatively few senior scientists/administrators. These scientists would often remain in their positions for decades. For example, Dr. Anthony Fauci has now served as head of the National Institute of Allergy and Infectious Diseases (NIAID) since 1984, a timespan longer than the corrupt J. Edgar Hoover spent heading the Federal Bureau of Investigation (FBI).

The Special Viral Cancer Program funds were used to support industrial laboratories, which were then used by NCI scientists for their own research. In addition, the scientists were often allowed to hire "outside contractors," essentially scientists for hire at private companies, to perform the work. My wife Sandy was technically employed by one of these private companies (Meloy Labs), while my work for Gallo was as an employee of Litton Bionetics. In addition to allowing principal investigators to hire many more people (because they wouldn't show up on the budget of the federal government), they were also able to pay the employees of these off-site contractors less money, give them fewer benefits, and make it easier to fire them. In essence, the legislation allowed people like Gallo to become the unquestioned ruler of a scientific empire, while at the same time transforming an entire class of researchers into the equivalent of medieval serfs.

Gallo would often command his contract employees to attend dinners where extramural scientists from around the world were wined and dined. The contract employees would pay the dinner expenses and get the company contractor to reimburse them because sending the bill to the federal government would have been an ethics violation. But Gallo, unlike most other NCI investigators, who did not engage in such activities, was somehow able to get away with it.

During the early 1970s, the Special Viral Cancer Program's single-minded approach to discover a human cancer virus led to several possible candidates obtained from continuously growing human cell lines. These cell lines were usually developed from cancer tumors, and it was believed that a viral infection is what had caused the cells to become cancerous, or essentially "immortal," and continue to grow, rather than be replaced, die, and be degraded, as happens with normal cells. These "discoveries" were usually announced in the media before peer review, and often turned out to be a mouse, cat, or monkey retrovirus which had somehow found its way into the culture, most likely from the addition of some product made from these animals included in the mixture.

When Gallo had wooed me at the American Society of Hematology meeting in December 1974, it was with the promise I'd come and work on a recently identified viral candidate, HL-23.[2] Scientists from around the world like Robin Weiss came to work on this human rumor virus.[3] However, it was discovered that HL-23 contained not one, not two, but three monkey retroviruses. At the time, Gallo was luring me with the promise of HL-23; he already knew it contained three primate retroviruses, meaning it was essentially up to me to start over.

As an explanation, Gallo first said that the patient's cells were contaminated, which we showed to be false. His next theory was that it was a plot against him by somebody in the National Institutes of Health (NIH) to discredit him. In private discussions with me it was clear Gallo was paranoid about enemies in the NIH being out to get him, and he fervently believed somebody in his lab had slipped these viruses into his cell cultures. The lab workers he suspected were never fired for fear of revealing the truth. To this day, I have no explanation why the HL-23 paper from Gallo's lab has never been retracted.

However, I was starting to learn some valuable lessons about my new boss. First, I learned to never show him data that I was unsure of because he would rush it into publication. The lead author on the HL-23 paper told me he was stunned to see the first public mention of his work in the pages of the *Washington Post* prior to being able to confirm his work, the information no doubt leaked by Gallo himself. The whole human embryo cells used to obtain the growth factor for the cell-line that produced the HL-23 virus had been lost in a freezer accident the previous November, which Gallo neglected to mention to me in December, making it difficult, if not impossible, to reproduce the work.

Later in the 1970s, viral candidates from the labs of two well-respected scientists (Henry Kaplan and Werner Kristen) turned out to be primate viral dead ends, too. As a sign of Gallo's tyrannical personality, although HL-23 had turned out to be a bust as a human virus, several researchers were interested in studying the HL-23 virus and requested samples, a common scientific practice. Gallo told his workers to irradiate the cell cultures before sending them to these researchers, which would kill any viruses in the sample. I cannot vouch that this was done, but I know that Gallo made the request.

I started to wonder what the hell I'd gotten myself into. Gallo told me I should attempt to reisolate the whole embryo growth factor lost in a freezer accident. This was needed to grow HL-23 myeloid leukemic cells and to isolate the virus he knew was causing acute myeloid leukemia (AML). That's all the direction I ever received, and one could still hunt for AML-causing viruses in 2021 because one has still not been found. Not much of a mentor. Simply a boss.

As one can imagine, these repeated public failures by Gallo and others were usually announced in the pages of the *Washington Post* rather than in peer-reviewed journals. This made the leaders in science proclaim loudly and to whoever would listen that human retroviruses did not exist.

As one of the human virus hunters in the Gallo lab, I started to attend the RNA Tumor Virus meetings held every year at the Cold Springs Harbor Laboratories. For many this was almost a mecca, a holy place of science, but I couldn't help but note that in the 1920s it had been the center of the American eugenics movement. The eugenics movement was based on the idea of white racial superiority. Italians and Southern Europeans were on the fringe of acceptable races, and this movement justified restrictive immigration laws, forced sterilization edicts, and gave Hitler a lot of the ideas for his "Final Solution."

As I attended these meetings, I was besieged by scientists, including five former Nobel Prize winners, who all vied to tell me how useless it was to search for human retroviruses. Many of them told me I should have made a better career choice and gotten a job with a real scientist, not one who was chasing phantoms.

The RNA Tumor Virus meetings started on Thursday afternoon. Following Saturday's presentations, there was a banquet on Saturday night. The human retrovirus research talks were always scheduled on Sunday morning, when most of the researchers were nursing hangovers and the leaders in the field did not attend the Sunday sessions. I was shocked by the misogynistic language used by many of these powerful male scientists, as well as their belittlement of women in science and their sexual crudeness.

The men I'd worked with on the road crew in Massachusetts when I was seventeen, who'd visit a house of ill repute on their lunch hour while I waited outside, were absolute princes compared to these scientists. My father had told me I had a choice of working with my back or my brain.

But these learned men of science seemed to be thinking from a completely different part of their anatomy.

* * *

When I started working at Litton's two-story facility in downtown Bethesda, Maryland, I met in Bob Bassin's lab a French post-doctoral fellow, Françoise Barré-Sinoussi, who would later win a Nobel Prize with Luc Montagnier for the isolation of the HIV retrovirus, establishing for all time how devastating retroviruses can be to human health.

It's amazing to realize how so many of us who interacted together at the NCI in those early years would have our lives so dramatically intersect.

Under Gallo's directive I searched for a growth factor in thirty-six different whole human embryo cultures, but had no luck. When I arrived at the NCI, the enormous Gallo lab operation was divided between several major scientists, including Alan Wu, Bob Gallagher, and Dave Gillespie. It was Gallo's habit to give the same project to multiple individuals in different sections and wait to see what developed. Unbeknownst to me, the project of using whole human embryos as a source for growing myeloid leukemia cells had previously been given to Doris Morgan and Zaki Salahuddin. They also had failed in the project, and I was assigned to replace Salahuddin and get it done with no other direction.

One of the many sources chosen to look for such a cell growth factor was mitogen (antigen) activated human peripheral blood mononuclear cells (PBMCs), even though the cells gave no indication of having such a growth factor. Human bone marrow cells, used as a source of myeloid progenitors, were cultured under many variations, with repeated additions of supernatants from stimulated PBMCs needed to prevent cell death. Despite many attempts, both Doris and I found independently that the immature and mature myeloid elements rapidly died, leaving only lymphoid cells.

At the time, it was thought only virally transformed B lymphocytes and leukemic cells proliferated in suspension culture. People in the lab were disappointed, thinking we only had B-cells. But B-cells did not require additional factors to grow in suspension culture. I thought the lymphoid cultures had to be something else, maybe T-lymphocytes, which had never been grown in culture before. I demonstrated that the continuously growing cells were almost pure cultures of human T-lymphocytes and could be kept alive for a long period of time by repeated additions of the T-cell growth factor I'd helped discover, which was later called interleukin-2 or IL-2.

For the rest of my research career, this discovery would be both a source of intra and extra scientific politics. When I first presented the work in abstract form, Alan Wu, Doris Morgan's supervisor] Bob Gallagher, my supervisor; and Ray Kiefer were coauthors. When it came time to submit the manuscript, those three names disappeared, which in my opinion could only have been through the intervention of Robert Gallo. Why didn't it dawn on me that such theft of credit could also happen to me at the hands of my paranoid boss, Gallo, who believed agent provocateurs from the NIH were secretly slipping monkey viruses into his cultures to contaminate them? The faith that good science and rational thought would save me was so mistaken! The lunatics were running the asylum!

Doris Morgan was also always claiming that these cells would only grow in small test tubes, but I was having success in the larger culture flasks as well.

Today, we would have published the first papers on the discovery as equal contributors; that is, as co-first authors.[4] The first author is the author who conceives of the idea, designs, and carries out the majority of the experiments and drafts the manuscript. Under pressure, I acquiesced and Morgan was the first author on the first paper of the discovery, T-Cell Growth Factor, and I was first author on the second paper. Morgan's version is that she always deserved to be first author, when in fact like the rest she thought that the lymphoid cells were B-cells and would never have hypothesized and proved as I did that they were T-cells. In retrospect, I believe we should have been equal authors on both papers. Her next project was to purify IL-2, which she failed to do, and soon afterwards, Gallo let her go.

Much later, Gallo started a whispering campaign through surrogates that I was the one responsible for Morgan leaving the lab. Nothing could be further from the truth. There was only one boss in that lab, and it was Gallo.

Gallo was initially disappointed that my approach seemed to generate the wrong type of non-myeloid cells for cultivating human retroviruses. He told me not to work on T-cells, but since production of retroviruses needs growing cells in order to replicate, we continued to look at the T-cells.

Gallo's lack of appreciation for the serendipitous nature of science was unfortunate. Until 1960, most scientists considered the lymphocytes to be an unimportant end stage of cell development. In fact, they should be considered some of the most critical guardians of the immune system, and their proper functioning is essential to the health of an organism. Peter Nowell, using phytohemagglutinin (PHA) to cause red blood cells to stick together, reported that this approach stimulated cell division in some white blood cells. Several investigators reported limited growth and survival of these cultures using PHA.

After our publication demonstrated we had discovered IL-2 and its properties, others claimed the credit. However, we had reported TCGF/ IL-2 (T-Cell Growth Factor, or IL-2), as a factor capable of generating normal T-cells in large quantities and predicted that these factor-dependent normal T-cells would provide an excellent tool for molecular and immuno-logical studies of the growth and differentiation of T-cells. In addition, the classification of T-cell subtypes made possible by this discovery could also be utilized in the treatment of cancer patients. Our predictions were quite conservative, as IL-2 has made possible an entirely new area of scientific research.

From the years 1976 to 1982, well over fifty investigators told me they could not reproduce my findings with IL-2 and growing T-cells in large quantities, as we had reported. It's important to realize that the failure of the scientific community to reproduce the data of a particular scientist does not mean that first scientist was wrong.

More often than not, I've found the trouble lies in the inability or unwillingness of the second researcher to follow the procedures provided by the first researcher. I know this will sound shocking to the average member of the public, but it has been my common experience.

Let me state this as clearly as I can.

Scientists often do not follow the precise materials and methods from the published study and then disparage you when they fail because of their own biases and dogma. Inviting such scientists to the lab to observe the results first hand, or sending them reagents, often solved the problem.

At first, Gallo's disinterest in my findings allowed me to collaborate with others, which was a good thing. Dick Metzgar from Duke University and I demonstrated that the cell surface of growing T-cells was far different than that of non-growing T-cells.[5] This dispelled the notion widely held at the time, that T-cells did not have cell surface antigens to kill foreign cells and pathogens. You might even say the T-cells were figuratively bristling with armor and sharp spikes to destroy invaders. Immunological studies were initiated by Kendall Smith and Steve Gillis, in collaboration with me. We showed the availability of IL-2 made it possible to study cloned T-cells with a single antigenic specificity[6], giving unparalleled data as to their inner workings. Smith's lab made seminal observations into the biochemical mechanisms of IL-2 regulated growth.[7]

In the concluding line of the abstract, we wrote, "The ability to sustain differentiated antigen-specific T-effector cell in long-term culture may pro-vide a means for the study of both the mechanism and regulation of T-cell mediated immunity."[8] In my opinion, that was a dramatic understatement. It's interesting to note that the material I supplied to their lab was originally referred to as "Frankie's factor."

The discovery of IL-2 stimulated the development of the new discipline of cellular immunology. The studies of many investigators over four decades have led to an understanding of how to therapeutically use T-cells' recognition of antigens, as well as the identification and use of receptors for IL-2 and its antigens. As Steve Rosenberg and colleagues wonderfully demonstrated, IL-2 alone could result in "durable, complete, and apparently curative regressions in patients with metastatic melanoma and renal cancer."[9]

The concept that manipulation of the immune system could be effective has revolutionized cancer therapy. The discovery that IL-2 is necessary for the development, survival, and expansion of T regulatory cells that repress immune responses against our own tissues (auto-immunity) led to the concept that IL-2 is nothing less than the *Yin* and *Yang* of positive and negative immune responses. These discoveries have led to the tremendous increase in attempts to use immunotherapy for treatment of chronic diseases.

It was particularly gratifying that at the hundredth anniversary meeting of the American Association of Immunology in Honolulu, Hawaii in 2013, our report of the discovery of IL-2 was selected as the second most important paper their journal had published in the last century.[10]

Considering the word lymphocyte did not appear in the index of the 1958 *Journal of Immunology* demonstrates the tremendous importance of that research in the field of immunology and the IL-2 discovery that gave rise to the new field of cellular immunology.

Things were less successful back at my Gallo-mandated project of trying to grow more than fifty human leukemia biopsies, with me eventually stopping at HL-58. You can't say I didn't give it a good shot. One of the pleasures of working at the NCI during that time was meeting the brilliant, young clinical fellows (men and women), who often reminded me of the famous quote by Will Rogers that "It's great to be great, but it's greater to be human."

One of my favorite people from that time was Steve Collins, who wanted to continue trying to culture human leukemia biopsies. I'll be damned if in his second attempt, with HL-60, from a patient with acute promyelocytic leukemia (APL), he was successful. Nobody was happier for him than me and that single discovery launched him into an outstanding scientific career. Steve and I would often laugh at the twists of fate in life as we had lunch at the Pines of Rome restaurant, just off the NCI campus. One person stops pursuing an idea, the next person walks in without the weight of that failure, and *voila!* the world changes. In a series of manuscripts, we demonstrated that these immature and nonfunctional cells could be made mature and functional.[11]

Maybe there's a lesson in that for all of us.

With another researcher, Ted Breitman, Steve showed that retinoic acid, a metabolite of vitamin A, at low amounts, could also stimulate this maturation (a process called differentiation).[12] That set the stage for a curative treatment for APL from a biologic! The idea that one could cure a leukemia with differentiation to a stage where the cells no longer grew, rather than killing the cells, was highly controversial. In essence, we were turning these dangerous, life-threatening cancer cells back into normal cells, thus restoring balance to the body.

Successful clinical trials in China and France were needed before Steve's treatment plan became standard in the United States. Years later, Judy and I had a student in our lab who was diagnosed with APL. We watched in delight at her recovery when her doctor used the treatment my friend Steve Collins developed. It was gratifying to have played a small role in the development of the therapy. But far more important was that working together we could accomplish what neither of us might have accomplished alone. That is how I'd been trained by Lew and Dane. Much more can be accomplished with a collaborative approach, rather than the cutthroat, competitive approach of Gallo. I am humbled by the scope of many of these problems. People's lives are at stake in the success or failure of research. We must bring our very best selves and bring back selflessness to science.

In 2016, the journal *Blood* honored our work in an article about the research that Steve Collins, Bob Gallagher, and I had done as one of the highlights of their seventy years of publication.[13] It was ironic in retrospect to be honored in this fashion, as none of us wanted to write what we thought was a review article because we were all so busy at the time. Another oddity I must note is how many times over the years I've been told by notable investigators that it's risky to be the first to publish groundbreaking research. Instead, they claimed, it's better for one's career to be second or third to confirm a discovery, and write the first review article which puts the research findings into their lab's credit. This has always struck me as a stunningly selfish way of thinking. To me it's like saying, "Don't be like the Wright Brothers and build the first airplane." Instead, be the idiot standing on the sidelines saying, "Hey, I think that thing just flew; I think I should write about flying one." Taking credit for the hard work and risks of others to enrich one's self should have nothing to do with the discipline of science.

Yes, it can be more dangerous to be first, but it's only on the frontiers of knowledge that you have the chance to change people's lives.

* * *

As I think you can understand by now, it wasn't just science, but the camaraderie I felt with other researchers that made this such an emotionally satisfying time in my life.

After work, Steve Collins and I would often play tennis. In the beginning, I would usually win. Steve was a hard worker and perfected his game, such that later on, I struggled mightily to win a match. I'd give him pointers on how to improve his game and he'd do the same for me. Afterwards we'd often get a meal or just sit around and talk about things.

In comparison, once when I played tennis with my boss, Robert Gallo, I beat him 6-0, 6-0, 6-0, eighteen straight games in a row. Gallo insisted on continuing, though. Gallo won the last two games. The final total was eighteen games for me and two for Gallo.

But that didn't stop Gallo from claiming afterward, "You see, I am a better player than you, Frank. I just needed longer to warm up."

I must note, purely for historic reasons, of course, that during the match, Gallo appointed himself referee, calling balls in or outside of the lines on both his side of the court and mine.

We never played again.

Another charming aspect of Gallo's personality was his treatment of his employees when they interviewed for another job. At one point, Steve applied for a position at the medical department of the Veterans Administration in Seattle, Washington. Gallo wrote a glowing letter of recommendation for Steve and then followed up with a phone call to the chief of staff of the department, denigrating Steve's abilities. For good measure, he added that all of Steve's good ideas had come from Gallo. In complete confusion, the chief of staff called me to get a better fix on what was really going on. I explained Gallo's behavior to him as best I could.

I guess I was successful.

Steve got the job.

I always enjoyed our two-way discussion about hematopoiesis and was honored when I had the opportunity to be one of his reviewers every five years when there was an assessment of his lab.

One of the last communications I had with Steve was several years ago, after he'd been diagnosed with a pancreatic neuroendocrine tumor. Steve had decided on a therapy protocol from the Zentral Clinic in Bad Berka, Germany. The treatment utilized radiolabeled peptides to bind and kill tumor cells, a procedure that was not available in the United States until recently. I remember thinking that if a brilliant physician like Steve couldn't find the best treatment in the United States, what chance did the rest of us have?

I dearly miss my conversations with Steve but am grateful for the years of friendship we shared.

* * *

Through the last half of the 1970s, nearly all virologists quit trying to isolate a disease-causing human retrovirus. The few stalwart souls who continued to make this effort were belittled by the leaders in the field who believed if they couldn't find such a virus, it didn't exist.

Fortunately, some good luck intervened. I've always loved the stoic concept of *amor fati*, which translates roughly as having the mindset to love fate, and make the best of what happens. The second of the three wonderful scientists I want to discuss in this chapter is Bernie Poiesz, a city boy like me, but from Philadelphia. Bernie had studied in the lab of Larry Loeb, where he learned how to test for reverse transcriptase, the enzyme used by retroviruses to convert their viral RNA into DNA, so that they could integrate their genes into the host cell genome. Bernie had just completed a year of clinical work and we were keenly interested in studying T-cell lymphomas (tumors).

Despite Gallo's instructions not to study lymphoid cancers, we knew that most retroviruses reproduced to detectable levels only in growing cells, which had been found in animals. That meant T-cell tumors, which grew quickly, would be an ideal candidate to find such retroviruses if they existed in humans.

While Bernie was demonstrating that IL-2 stimulated growth of these tumor cells, I reviewed a paper by Adi Gazdar, which reported that IL-2 had been successful in establishing several human tumor T-cell lines.[14] Bernie knew Adi and his boss from clinical ward rounds and had been given several of these T-cell tumor lines in a collaboration. In a marvelous irony, the most interesting cell line, Hut-102, was developed from a patient Bernie had admitted for a lymph node biopsy. Bernie found small, but measureable levels of reverse transcriptase in the sample.

We were encouraged, but also disappointed because we wanted to find extraordinary amounts of reverse transcriptase, hoping to show a great amount of viral replication as well as a virus to convince the skeptics.

Bernie was wonderfully precise in developing an experimental plan, and we developed an "A-team" of all the researchers in my lab to accomplish the Herculean task of producing thirty-two gallons (120 liters) of cell culture for viral analysis. Bernie was then able to identify a retrovirus.

We knew we had something, but what was it?

We were concerned about a rerun of the HL-23 fiasco, where the three viruses isolated were later found to have come from a monkey. Bernie knew that Adi Gazdar often passaged his human cells through mouse tissue and worried the virus we had might be a mouse virus. Bernie informed Adi of our results and our concerns about the meaning of the data, and Adi generously provided another sample of which there could be no question of mouse contamination. In a painfully slow process, we checked all the available retroviruses and confirmed no animal retroviruses were in the sample.

The data suggested we had isolated the first known disease-causing human retrovirus. We had reverse transcriptase, we had the virus, and we'd compared it to known animal retroviruses and had come up empty.

At this point, we had not told Gallo of our findings because we wanted to check our work before we found it splashed on the pages of the *Washington Post*. However, Adi Gazdar's boss, John Minna, apparently loved to one-up Gallo, and told him of our discovery when the two of them were at a Saturday night cocktail party at Gallo's home.

Both Bernie and I were awakened late that night by Gallo screaming at us over the phone about why we'd kept this information from him.

At the tense Monday morning staff meeting that followed, Gallo was still angry at us and felt the need to reassert his authority. He asked us to provide him a readout of the reverse transcriptase counts (which we had ready), and after reviewing them declared in an imperial voice, "I know viruses and you don't have a virus. Go back and prove it."

Gallo then insisted on being in charge of the project, with regular meetings with the three of us. However, not only were the findings about Hut-102 reproducible, but we were able to establish the presence of the retrovirus in uncultured cells of T-cell tumors. We were also able to establish the virus was only present in T-cells and not B-cells of a second patient.

Bernie and I had bagged the first identified disease-causing human retrovirus.

But of course, it was Gallo who wanted to claim all the credit.

* * *

While the science was going well, the politics were not.

In keeping with tradition, Gallo was given the right to name the retrovirus. When he brought it up to us, we encouraged him not to be like Rous, Maloney, Friend, and Epstein-Barr, and give the virus the name of a person. Gallo finally agreed and named it the Human T-Cell Lymphoma Virus (HTLV-1). Bernie properly wanted to be cautious as we did not know for sure if it was of human origin and whether it caused disease. The evidence certainly suggested these two ideas, but there had not been rigorous investigation.

In publishing the first paper on a new infectious entity, disease causality is almost never proven. Gallo put us in a difficult position because those critics who said that human retroviruses did not exist were now claiming we were asserting this was a disease-causing virus. However, Gallo was adamant about the title. He was going to plant his flag on this virus. We were going to have enough baggage associated with our discovery without the continued sniping by our critics. Even in our lab, other scientists who had not studied HTLV-1 because they did not believe us were now resentful for being left out. However, some researchers did come to our side, and we welcomed their assistance because it was clear that many important questions needed to be answered.

There were a number of odd events that took place as we continued our investigation. One day, Bernie's lab book mysteriously disappeared. Bernie eventually found it on the desk of Mika Popovic, a new virologist from behind the Iron Curtain who was working at the NCI. Remember, our lab at Litton Bionetics was miles away. Gallo defended Popovic against Bernie's concerns, but Bernie thought the defense insincere. He believed Gallo had, in fact, set up Popovic to take the blame for Gallo's removal of Bernie's lab notebook from our lab.

While preparing the manuscript, Gallo asked the editor-in-chief of the *Journal of Virology* to review the paper. Gallo reported to us that the editor replied, "No. And you should stop working on these viruses. We know they do not exist."

In the process of submitting the paper to the *Proceedings of the National Academy of Sciences* (*PNAS*), where it was eventually published in December 1980, Gallo told me he wanted to be first author on the paper.[15]

I replied that Bernie had done most of the intellectual and technical work and deserved to be first author.

Gallo looked out the window, and after a moment said, "Frank, with an attitude like that you're never going to get ahead in science."

As I walked out of that office, I knew my future working for Gallo was bleak.

* * *

The *Proceedings of the National Academy of Sciences* paper was not well-received, especially since HTLV-1 had not yet been definitively associated with human disease.

The next May, in 1981, Bernie and I were each given short five-minute talks to discuss our research at the Cold Springs Harbor RNA Tumor Virus meeting—of course, on a Sunday afternoon. From the start of the meeting on a Thursday, until Saturday, Gallo had us call him every day and deliver our presentation to him. He'd suggest minor changes and then ended each conversation promising he'd be there for our Sunday afternoon talk. Needless to say, he didn't have the courage to show. After our talk to a half-empty room, everybody filed out without talking to us, with the exception of future Nobel Prize winner Harold Varmus. He came up to us and said, "I think you have something there. Keep working on it."

In the spring of 1981, Gallo flew to Japan to meet with scientists studying acute T-cell leukemia. Drs. Hinuma, Miyoshi, and Ito did excellent work identifying retroviral particles in cells with the disease. This was critical in establishing the virus caused the disease, but Gallo was always dismissive of their contribution.

Our work and that of many others clearly demonstrated HTLV-1 was the first identified human infectious cancer-causing retrovirus, which in addition to lymphoma, also caused an accompanying neurological disease, called Tropical Spastic Paraparesis/HTLV-1 associated myelopathy, TSP/HAM. At the time, we knew this was revolutionary, but did not realize the major impact this discovery would have on methods used for viral investigation and the resulting therapies.

For its one hundredth anniversary, the *Proceedings of the National Academy of Sciences (PNAS)* selected our paper as one of its thirty most important papers. John Coffin, never a particular friend to me or Judy, and the most vociferous critic of our future XMRV work, wrote of our HTLV-1 paper:

> There is little doubt that, without the trail blazed by the highlighted PNAS papers and its predecessors, the discovery of HIV which led to screening technologies to prevent transmission and eventually to effective antiviral therapy now prolonging the life of millions of infected individuals worldwide would have been greatly delayed. Contributions to delays were the pre-1980 mindset among virologists of the unlikelihood of human retroviruses. Thus, both the technology of growing T-cells in culture and isolating viruses from them and the paradigm represented the discovery of a pathogenic human retrovirus,

laid the critical foundation for some of the most important medical advances of the late 20th century, without which many more millions would have died of AIDS.[16]

It's a remarkable statement because it shows how negative dogma proposed by arrogant and powerful leaders can hurt public health, and how many of the current critics of our work have no problem admitting the importance of our work in the past.

In retrospect, the potential of long-term growth of T-cells, and subsequently other cell types, including immature myeloid cells, the original cellular target, was as important as the discovery of a single growth factor or virus. For example, we showed that stem cells from human cord blood cells could support cell growth and differentiation in long term cell cultures, which suggested the stem cells from cord blood could be used medically.[17]

This was another paradigm shifting discovery, years before its acceptance by the gatekeepers of medical dogma.

* * *

The last of my clever and humane clinical fellows I wanted to talk about is Joe Gootenberg, a progressive thinker who first made me aware of the deplorable medical care on Native American reservations.

Joe and I first worked together demonstrating these T-cell tumor cells had growth advantages because they produce their own growth factors.[18] Using our old friend, the Hut-102 cells, Joe found a biochemical variant T-Cell Growth Factor, and we presented evidence that they were different molecules and could be expressed by separate genes.[19]

In 1984, Joe set up his own lab at Georgetown University and submitted several grants to isolate the gene for this new variant T-Cell Growth Factor. His research was not funded because he was not a member of the club and did not have experience with the new techniques. How does one get experience without getting a chance? Ironically, the gene for the new T-Cell Growth Factor which we had discovered in 1984 was cloned, sequenced, and named IL-15, by the NCI in 1994 and a company named Immunex, where investigators did not have to apply for grant money. They were not allowed to patent their discovery because of our prior knowledge.

Ten years into my career, it had been my good fortune to have contributed the knowledge of several critical discoveries to medical science, which had been decorated by highly honored publications. The discoveries fostered therapies which have prolonged lives and improved quality of life for many individuals. In addition, there were also anti-IL-5 therapies for severe asthma and other diseases with eosinophil pathology. IL-15 has already shown some promise in expanding the range of cancers treatable by immunotherapy.

But why didn't I have a sense of great satisfaction to be present at these moments of discovery? I should have been happy knowing what I had achieved, and indifferent to the fact I was unlikely to get much external acknowledgment.

The problem was that I was working for a narcissist peacock, who had an exaggerated sense of entitlement and required constant complete admiration. He was reckless, doing things like prematurely announcing the discovery of HL-23 in the pages of the *Washington Post*.

It was like watching a madman play with a pistol.

You never knew who was going to get shot.

He kept us waiting every Monday morning staff meeting beyond the time the meeting was supposed to start because as the boss he believed he was the most important person in any room he might occupy. I was told I should not ask challenging questions of other scientists who visited our lab for fear it might make Gallo look bad politically. I was told not to discuss my work with my wife for fear she'd tell her bosses, who would then steal Gallo's work.

Science was supposed to be about collaboration, especially since as government scientists we worked for the public, but not with egomaniacs like Gallo and Fauci. It was all about their own personal glory.

Gallo had no idea how little other investigators at the NCI thought of him. Gallo would constantly discredit any information which contradicted his pre-conceived ideas or the possibility he was not the center of the scientific universe. It took me a while to realize Gallo actually believed his own lies, even though he kept changing his story. I was amazed at how his mind could so consistently rewrite reality and still believe the new lies he told. Thus, in Gallo's mind he *was* the better tennis player, all the good ideas *did* come from him, and he *should* have been the first author on the HTLV-1 paper, not Bernie Poiesz.

It's been widely reported that I had many verbal battles with Gallo. This is true, but those came after I left his lab, and after he continually tried to destroy my ability to work. I never publicly contradicted him while working for him because I had seen senior leaders disappear if they crossed him.

Maybe I should have, but I'll leave that decision to others.

Gallo would often come back from reviewing grants for a scientific society and at a staff meeting, pass out their grants for us to read. I was horrified by this amoral behavior. Gallo was encouraging us to steal these ideas from scientists who'd submitted their research proposals with the expectation they'd be kept confidential and that their leaders were ethical.

I always replied, "Thanks, Bob, but I've got so many of my own ideas to pursue that I don't think I'll be able to get around to any others."

I thought it was my job to do good science for the public, despite the negative behaviors of my boss. I believed the good which I could do would offset the negative parts of Gallo's personality and behavior.

Boy, I was wrong.

CHAPTER FOUR

Political Unemployment and Career Threats

The complex pattern of the misallocation of credit for scientific works can be described as the "Matthew" effect. The Gospel says, "unto everyone that hath, shall be given and have abundance, but for him that hath not, shall be taken away even that hath." Thus, the Matthew effect consists of the accruing of greater increments of recognition for scientific contributions to scientists of considerable repute and power and the withholding of such recognition from scientists who have not made their mark.

—Robert Merton[1]

As soon as Gallo believed it had been convincingly demonstrated that HTLV-I, discovered by Bernie Poiesz and me, was a disease-causing retrovirus in human beings, he put into motion his propaganda machine to convince the scientific world that his version of the events leading up to the discovery was the Gospel truth and all others were the most unspeakable of heresies.

For example, Gallo always claimed credit for proving the link between HTLV-I and ATL, which was not even close to the truth. That work had been done by Drs. Hinuma and Miyoshi of Japan. Gallo often said publicly that, "most of the work from Japan is not terribly relevant," a lie which was clear to anybody with even a passing interest in the field.

The discovery of HTLV-1 and my work with IL-2 gave him the opportunity to claim dominance in the field. Gallo once sent me to a meeting with our NCI collaborators who had helped develop the cell lines we'd used to isolate HTLV-1 to gently tell them their involvement in any further studies would be marginalized. Gallo wanted to hoard HTLV-1 and all retroviral studies to himself the way King Midas wanted all the world's gold. He desperately needed sole recognition for this discovery and any efforts moving forward.

There was one instance when I needed to see him on an important matter but could not get a meeting with him for several days. I wondered if he was dodging me, only to find out later he'd been at home for several days, in apparent mourning, because he had once again been denied membership to the National Academy of Sciences. When I finally did get to meet with him, all he could talk about was how his enemies had screwed him again.

During the HTLV-1 discovery process, it became clear how critical my work on IL-2 was for the field. He quickly published a review on IL-2, then demanded to be a coauthor on all my future IL-2 papers. This demand was given in the middle of him not speaking to me for six weeks and then shouting at me because a French scientist had asked if he was still working with me on IL-2.

At other times he tried to ingratiate himself with me and pull me into his web of fantasy. One day he told me in private that the two of us could be like the famous scientific partnership of Harold Varmus and J. Michael Bishop, who would go on to win the Nobel Prize in Medicine in 1989 for their research on the origins of cancer. I knew those two researchers each had their own resources and political power base, while I didn't have either of those. In essence, Gallo would be the high school bully trying to get the competent kid to do his homework.

Before Bernie Poiesz came to Gallo's laboratory, he already had a position in Hematology-Oncology at the State University of New York (SUNY) Medical Center in Syracuse. Bernie returned to Syracuse after his time in the Gallo lab expired, but he didn't escape unscathed from Gallo's wrath. While Bernie didn't need Gallo's help to get a job, since SUNY wanted him back when he was finished with his work at the NCI, Bernie needed a recommendation letter from Gallo to receive funding to continue his HTLV-1 studies.

How did Gallo respond?

He wrote letters to the agencies which would have provided grants to Bernie, demeaning the quality of Bernie's work. Bernie responded by threatening to sue Gallo, who then backed off.

Shortly thereafter, I received a handwritten memo from Gallo, via the resource manager, about how I would not receive any more leukemia samples or viral preps, thus cutting me off form further HTLV-1 work in the field I had help create.

I'm sorry to report that Gallo's methods worked, because in the years to come whenever people talked about HTLV-1 it was always Gallo, Gallo, Gallo. Years later Bernie said to me, "I can't change how people perceive it or how people present it in the media. The only thing I can do is my work. I spent many, many nights in the laboratory. The moment of discovery was ours, not Gallo's."

Over the years I have struggled to match Bernie's positive attitude.

* * *

Out of the blue, Gallo called me from his office once to tell me he'd just talked to a perfect post-doctoral candidate for me.

"What does she study?" I asked.

"I can't remember," he replied.

"How will I know her?" I questioned.

"She has the best set of boobs I've ever seen," he replied, then broke down into a head-shaking laugh.

Dr. Christine Easement turned out to be an insightful, passionate investigator, with a first-class mind and the temperament to pursue challenging questions. As a competitive swimmer, she missed the Olympics by less than one hundredth of a second. She had developed a hamster bone marrow cell culture to study blood development from stem cells. Previous dogma said these cells had to stick to the bottom of the culture cell flash. Hers did not, and the collective wisdom was that her experiments must have failed. They had not, but it took us collaborating on several papers before one got published and the prevailing dogma was overturned.

Of course, Gallo was not interested in this work, simply happy that I'd been eliminated from further work on HTLV-1.

But my time with Gallo was growing short.

I just didn't know it.

* * *

By 1981, I felt my scientific career was in a straitjacket. Gallo was keeping me off the challenging projects because I was getting too much credit. Instead,

I was looking for work that might be important, while at the same time try-ing to convince Gallo that what I was doing probably wouldn't lead to any groundbreaking discoveries.

To make matters worse, in September 1981, my father passed away of Hodgkin's Lymphoma, telling me with his last breath that he was disappointed because I hadn't given him a male heir. It was gut wrench-ing, because I felt if I had been more attentive, his life could have been prolonged.

Meanwhile, several scientific friends of mine who shall remain nameless for fear of retaliation said Gallo's resurgence was my fault because Gallo's career had been in the dumpster before my discoveries had saved him.

That one hurt because it had some truth to it.

The final blow came around my birthday in February 1982 when I received a letter from Gallo informing me that my position as a cancer expert would not be renewed, and by May I'd be out of a job. In the letter, he also advised me not to limit my job search to the Washington, DC area.

The message was exceedingly clear.

Gallo wanted me gone from his sight. When I went to his office and confronted him as to why he was letting me go, I was shocked by the hon-esty of his answer.

"You're getting too much credit," he said.

In what universe was I getting too much credit?

I had hoped the scientists who mattered knew the phoniness of Gallo's claims.

Sandy and I understood that for each of us to obtain a position com-mensurate with our talents and experience, in the same city, would be a challenge. However, there were options. In 1978, Kendall Smith had tried to recruit both of us to Dartmouth Medical School in Hanover, New Hampshire. That qualified as being out of the Washington, DC area.

Although both of us had always wanted to teach at a university, the attractions of academic life were not what we'd expected. Her recruiters told us that Sandy could be the first woman in the department to receive tenure there, something which immediately set off alarm bells in our heads. Ten years later, when I gave a presentation at Dartmouth, the female research fellows told me the situation remained the same.

No woman had received tenure in the department.

My recruiters also told us that if I was hired, I'd be the only person in my department who hadn't graduated from Harvard. Nothing like a private little boys' club, is there?

Before we even decided, Steve Collins sent me a copy of the Dartmouth alumni journal with an item that announced the two of us were joining the faculty. The galling arrogance of the place had turned us off. It's difficult even now for me to believe that despite the misogyny we witnessed in graduate school and at the NIH, the NIH still provided more opportunities for women than most universities. For today's situation, see the documentary, *Picture a Scientist*.

Our experience at Dartmouth in 1978, although disturbing, would end up being my most positive experience in the job search. I answered an employment ad in *Science* for the new armed forces medical school which had several faculty positions. I phoned the chairman of the microbiology department, who, after looking at my credentials for a few minutes, said, "You are the best candidate for this job. Please call me back when the ad closes."

When I did, he claimed he'd never talked to me! What do you say to that? Had Gallo gotten to him?

A position at the armed forces radiologic institute also mysteriously disappeared.

When I talked to many of the famous and well-respected scientists with whom I'd worked, and who I knew disliked Gallo as much as I did, they told me nobody would hire me because of my association with Gallo. What was deeply disturbing was how many of my former colleagues refused to write me letters of recommendation because of their fear of Gallo's wrath. One time, I was talking to a scientist in Seattle, asked for the letter, and actually heard him close his office door before asking if I thought Gallo would approve of him writing such a letter.

I replied that he'd probably have to ask Gallo himself, but my guess would be no.

He declined to write me a letter.

A friend from the Special Viral Cancer Program offered Sandy and me a job with a company in Seattle for less money and informed us we'd have to share a single technician. What an insult. That was definitely not a job we would consider.

Without going into all the odd and terrible experiences of that time, let me simply say I got nowhere for six months. As a result of that experience, I have realized how thoughtless it can be to tell a recently unemployed friend that they'll have no trouble finding a job. As each day passed without success, I sank deeper and deeper into depression.

An amusing incident occurred while I was unemployed, which perhaps was the universe telling me that even in the midst of my misery, my position had substantially improved. One Sunday morning I received an unexpected call from Mary Jane Gallo, the wife of my former boss. From my years of working with Gallo I knew he had a roving eye. One time, Gallo lost one of his attempted paramours to another young scientist and must have waited for the appropriate moment to call their room, then hung up when they answered.

The next morning, I had breakfast with the scientist, and he asked if I'd been the one to call the room at Gallo's direction.

I told him I had not. It must have been Gallo. That was his vengeful style.

Mary Jane Gallo's call to me that morning was not to complain about her husband's wandering eye, but to tell me that he was buying this particular woman in the lab expensive gifts with his Lasker Prize money, which had been given to him for my HTLV-1 research! According to Mary Jane, she'd told Bob to get rid of the woman, but he claimed she was too important to the lab.

I said to her, "Think of all the people you've met, who've now disappeared from his lab. Your husband does not need anyone. It's just a ploy to keep her around the lab."

Mary Jane must have listened to me because that woman soon left the lab for an endowed chair at the University of California, San Diego, more than two thousand miles away! Such are the wages of sin!

I guess it was a pyrrhic victory for Mary Jane, but I wonder if she gained anything of lasting value. I have never understood the mistreatment of women by powerful men but am equally confused by the women who choose to remain at the sides of such men.

* * *

Meanwhile, my career was at a crossroads. Every day I was unemployed I felt like an athlete whose skills were diminishing each moment I was off the field.

I had an excellent interview with Dr. Ruth Kirschstein, a wise, warm, and caring director of the NIH's General Medical Science Institute. After reviewing my application, she said, "Boy! I would love to have your resume!" She kindly offered me a permanent civil service position to be a program administrator. It also came with a nice bump in salary from my position in the Gallo lab.

But it came with a catch.

She said, "Frank, you have to decide whether you want to be a referee or a player."

Although I hated to turn her down, I knew I could only be a player, meaning my destiny was to be a researcher in the lab.

As part of his 1971 War on Cancer Act, President Nixon requested that the US Army transfer land and buildings which had previously been used for the biological and chemical warfare research program at Fort Detrick, Maryland, to the NCI for use in cancer research.

When Reagan came into office in 1981, he greatly expanded the use of these outside contractors, firing many scientists, but then bringing them back under the umbrella of these corporate, for-profit labs. Of course, the researchers were paid less, had fewer benefits, and the intellectual freedom to stand up to powerful governmental or corporate interests was greatly reduced.

However, the decision to turn Fort Detrick from a biological and chemical warfare facility into a cancer research program did have some benefits for me.

One of the programs they were setting up at Fort Detrick was an innovative program called the Biological Response Modifiers Program (BRMP). The impetus for the program had been the twin research for interferon, a naturally occurring biological molecule which attacked both viruses and cancer, and IL-2, which the new laboratory chief, Dr. Ron Herberman, had discovered activated a natural killer cell that attacked cancer cells. Herberman offered me a position to study these immunological mechanisms at the new NCI on the grounds of Fort Detrick, and I eagerly accepted. There was also the possibility of a good position for Sandy there as well. Despite our fantasies of academia, it was hard to leave the NCI.

It had been a long six months, but as Labor Day, 1982, approached, Sandy and I went to Boston to visit with my mother and siblings. I lost a vigorous game of tennis with my brother, then when I returned home my mother said Gallo had called and left a return number. My heart sunk.

I nervously dialed, wondering what fresh troubles might arise, but when I reached him, Gallo said, "I'm sorry you can't find a job. But I have a position for you in my lab, Frank."

I paused for a moment, then said, "Thanks, Bob, but I have a job." I was feeling pretty good about my position. But Gallo, the master chess player, had a move I hadn't anticipated.

Gallo replied, "Haven't you heard, Frank? It's been eliminated."

When I returned to Bethesda, I called Herberman and asked what the hell had happened. Herberman replied, "Frank, if I were you, I'd get a lawyer. Gallo convinced the director of cancer treatment to eliminate your position."

I knew Gallo had been getting criticism from his colleagues about how I'd been treated by him, although he claimed he'd never received a single phone call from anybody about me for a recommendation. Gallo made me many promises of work freedom, which I knew were lies, but I reasoned it was easier to get a job if I already had a job. But in my mind, I knew Gallo's approach. He'd say he gave me a second chance; I couldn't do anything, couldn't get a job anywhere else, and he'd eventually have to let me go.

It was a second chance to walk the plank, but now I knew what I was doing.

After twenty weeks of unemployment and feeling completely useless, I reported to the Gallo lab again in October 1982. Despite receiving the outward trappings of a valued scientist, a small lab and an office, I wasn't given anything to do. I wasn't allowed to attend staff meetings and the lab members were given strict orders not to talk to me, which I'm sorry to report, most of them obeyed.

I was in Gulag Gallo, silenced by the salary I was getting, rather than being needy and complaining on the outside. I considered Herberman's advice to get an attorney, gather my evidence, and prepare for the inevitable confrontation with Gallo.

In addition, I studied the latest literature and kept my ears open for any other positions.

Over the years, many people in and out of science have asked me how the people involved could have let this happen.

The simplest answer I can give them is that it's all about money needed for career advancement.

Most scientists will justify any action that provides them access to research funds. Unfortunately, scientific research has a lot of bad actors, petty tyrants, and mediocrities who are desperate for funds. The convergence of the SVCP and Nixon's "War on Cancer" meant that an unprecedented amount of research money was available, without the traditional controls. That allowed for the development of many enormous empires masquerading as independent scientific research laboratories; when in reality, they danced to the tune of the chief scientist.

In the early 1980s, Gallo's lab consisted of fifty scientists and had an annual budget of thirteen million dollars. (In 2021 dollars, that's equal to about thirty-six million dollars.) Naturally, every overseer, such as the secretary of Health and Human Services, would chant the mantra that "the Gallo lab is doing cutting-edge research," in order to justify the vast amount of money being spent. It was like admirals and generals showing up to testify in front of Congress that the money was being spent to "protect the homeland," and all the spineless Congressional members couldn't open the public checkbook fast enough.

The growth of these mega-labs continues to be a problem today. In 2012, Ian Lipkin, a professor at Columbia, told me that with sixty-five people in his lab he could not compete with other labs. Sixty-five researchers and you don't feel like you can compete? Like the robber barons of old, competition to these scientists is the absence of competition.

Another key issue was the vindictiveness of certain individuals, especially toward those who might criticize them. When *Science* magazine published a history of the Special Cancer Viral Program, a footnote explained that most refused to have their names published for fear of reprisal.[2]

Since people are afraid of the consequences of telling the truth, only the fairy tale praise of the sycophants makes it into the history books. Media pundits, with no idea of how paradigm-changing work in science is accomplished—the teamwork and enormous work necessary—only make matters worse by their hero worship of unworthy individuals like Gallo and Fauci who claim credit for the hard work of others. Except when that work reveals inconvenient truths of the damage wrought by the scientific establishment. Then the misogynists resort to the traditional cowardice of blaming the messenger, especially if that scientist is a woman.

Furthermore, the unremitting servitude of the students and postdoctoral fellows allows the established principal investigators like Robert Gallo, Anthony Fauci, and Ian Lipkin to take credit for the work of their underlings like the masters of a vast, intellectual plantation. Post-doctoral fellows are keenly aware that if they don't leave places like the NCI, Harvard, or wherever they do their work, in the good graces of the powerful, their career has no chance of survival. It is extremely difficult to be an honest person in this type of environment, especially when such honesty is often likely to get you a one-way trip out of your chosen field.

Finally, empire-building fosters the development of huge bibliographies, usually of the principal scientist in charge, to justify the enormous budget. The claims of some of these scientists rival the honors all-powerful dictators often give themselves, such as single-handedly diverting the course of mighty rivers. Gallo now claims to have more than twelve hundred publications to his name.

If we assume Gallo had a forty-year productive career, that means he published a new paper every twelve days. It's preposterous to pretend that anybody can design, conduct, and publish that many important or original papers. Many of his "publications" are simply opinions or commentaries, not much better than a typical blog post, which he used to deflect, deny, or spin the truth to his advantage.

In early 1983, the Biological Response Modifiers Program (BRMP) published an ad that it was going to advertise for the positions which had been eliminated the previous September. (Remember, nothing ever dies in a bureaucracy, but it can be put into a deep sleep.) Figuring I had nothing left to lose, I put my application in again.

One day I received a call from a more senior scientist, Dr. Joost J. Oppenheim, who had applied for a section chief position. Joost was an immunologist at the Dental Institute and invited me to lunch. He'd heard stories of the vindictiveness of the leaders of the NCI and knew they did not look kindly on the Biological Response Modifiers Program.

I told him that the Fort Detrick facility in Frederick, Maryland was looked upon by the NCI powerful elite as akin to being transferred to Siberia, and they might leave a researcher in relative peace to do some actual science. I suggested that as a new institution, the political divisions and turf battles might not be as severe as in a more established facility.

The lunch was a turning point in my life because Joost got the section chief job, then hired me to work in the BRMP, just as I'd hoped to do the previous September.

Years later I asked Joost if Gallo ever harassed him for hiring me.

He replied that Gallo called him repeatedly and threatened him not to hire me.

When I asked why he didn't give in to the pressure, Joost laughed.

"Because I couldn't find anybody cheaper!" was his response.

And besides, an acquaintance of Joost's, Steve Gillis, later founder of a successful biotech company called Immunex, told him, "Give Ruscetti a hundred thousand dollars, lock him in a lab, and he'll discover something!"

At that time, a hundred thousand dollars was cheap for producing ground-breaking science. I always felt that was a nice compliment about the quality of my work and dedication.

Joost Oppenheim was a male Anne Frank, forced as a child to hide in Holland from the Nazis during the Holocaust. I think that experience made him the kind of person who didn't have trouble standing up to a bully like Gallo. As a sign of our friendship, a group of us, headed by Howard Young, gave Joost a copy of the diary he kept during those years, translated into English. I shall be forever thankful for Jo's generosity and integrity.

It was a wonderful relief to stay in the area, because Sandy had also gotten a job at the NCI, and she was having great success in her research highlighting the pathogenic changes to cells caused by retroviral surface (envelope) proteins, which in some viruses included spike proteins.

As the spring of 1983 dawned, I was well-aware that there were many villains in science, but was discovering that there were several heroes as well.

CHAPTER FIVE

The Mischief Behind the
Discovery of HIV and the Rise
of Anthony Fauci

Facts do not cease to exist because they are ignored.
—Aldous Huxley[1]

Be innocent of the knowledge, dearest chuck; TILL thou applaud the deed.
—William Shakespeare[2]

Sandy and I were delighted that my new job allowed us to stay in our condominium in Washington, DC. Her commute to the main NIH campus in Bethesda, Maryland for her job wasn't more than ten minutes, while I had to travel another thirty-seven miles down the road to the Fort Detrick army base in Frederick, Maryland, which was then being referred to as the Frederick Cancer Research Facility (FCRF).

Sandy had joined a genetics laboratory supervised by Michael Potter, one of the most humble and unaffected scientists I'd ever met, who was also very accomplished in his field. He restored my faith in a good man doing great science. Sandy's group was the first to demonstrate that the expression of a retroviral protein envelope alone could cause cancer in mice.[3] This envelope protein was part of a defective virus called "spleen-focus-forming-virus," a virus which had to be delivered to cells via another, infectious virus. In other words, in order to cause damage, the defective envelope had to be delivered to the target cell.

One should understand a virus is usually divided up into two parts, the so-called lipo-protein envelope "shell" and the interior of the virus which contains the genetic information, along with structural proteins. Researchers have long believed the "shell" of the virus was secondary to any disease process, claiming only the interior of the virus carried any disease-causing factors.

Sandy's research was turning that belief on its head, suggesting the shell of the virus might also be dangerous to living things. This was especially important in terms of the reigning dogma, then and now, that only infectious transmissible viruses can cause harm to an organism. I strongly believe that so-called "defective viruses," meaning they only have part of the full genetic code of the virus (possibly just a few hundred base pairs from the shell), can cause significant harm.

It can be difficult to visualize what I mean with our frame of reference, because when we see an organism, like a dog or a cat, we see the entire animal. We don't just see a tail, an ear, or a stray leg, making its way down the street. However, in the viral world, you might get merely a segment of the genetic code replicating. In the past, this would cause researchers to label that a "defective" virus or "partial viral strand," and they'd categorize it as harmless. Personally, our hypothesis is that some of today's chronic diseases may be exacerbated by these "defective viruses" that we actually see expressed aberrantly from what we call the endogenous virome (a group of viruses present in the animals), but do not classify as being of any pathologic importance.

In April 1983 when I started work at the Fort Detrick facility in Frederick, it had been nearly two years since I'd had the opportunity to engage in any meaningful research, and I was anxious to get started. The level of funding I'd receive was still uncertain, but I was given permission to hire one technician.

Of all the laboratories in the world, Judy Mikovits had to walk into mine.

* * *

Franklin Roosevelt once famously said "I ask you to judge me by the enemies I have made."

The fact that Judy's enemies and my enemies are usually the same people means you should probably judge us similarly, despite any differences we might have in personality, temperament, or even interpretation of the data.

As a woman in science, Judy has been fighting against sexism and misogyny since her high school science classes in ways that often stagger my imagination. I cannot fathom the behavior of some men. But Judy has never backed down.

After declining her acceptance to Princeton University for financial reasons, Judy attended the University of Virginia and graduated in 1980. It's worth noting that when Judy entered the School of Arts and Sciences in 1976, it was only the second class of women admitted in the 152 years since Thomas Jefferson had founded the University. Many of the professors were not supportive of women at the university and had little hesitation expressing their opinion. Judy learned as I had, years before, that professors and administrators often discharged their duties without regard to merit, but by whichever course of action was politically easier.

Less than a month after her graduation in May 1980, Judy was hired as a chemist technician for the Fermentation Chemistry Program at the Frederick Cancer Research Facility. Her first project was purifying interferon-alpha (IFN-a), which at the time was being investigated as a highly potent anti-cancer and anti-viral therapy. (Interferons are small proteins created by cells, primarily in response to viruses and bacteria, that can positively modulate the activity of the immune system.)

Both IFN-a and, later, IL-2 were inflicted with the same exaggerated expectations every new therapy receives. First, the new therapies are believed to be cures for multiple conditions. When they are used without application of knowledge and fail to be the magic bullet cure-all for these multiple conditions, they become unwanted dog shit, and struggle to find their useful therapeutic niche.

In 1982, the fermentation chemistry team for which Judy was working was assigned a project, taking her off the interferon-alpha project. Gallo wanted the contractor team to provide his lab with thirty grams of IL-2 and HTLV-1 virus from a two-hundred-liter (about fifty-two gallons) culture of Hut-102 cells, using a procedure developed by me and Bernie Poiesz. In order to extract the thirty grams of HTLV-1, the protocol required an open-air centrifuge, known as a K-centrifuge.

A centrifuge changed my life once before by getting me transferred from the Air Force to the space program.

Now, a centrifuge was going to bring Judy and I together.

Mark, the team leader, and Judy immediately realized that the use of the centrifuge had the potential to cause the virus to become airborne to infect the technicians, including a few who were pregnant. Little was known about the transmission of the first identified human disease-causing retrovirus, its infectivity, or the mechanisms through which it caused disease.

Mark and Judy wrote a letter to their superiors at the NCI requesting additional time and resources to complete the assignment with proper safeguards. The threatening reply from Robert Gallo was a simple edict to meet the deadline or lose their jobs. Additionally, they were not to reveal any of their concerns to the other lab workers. In Judy's naiveté, she even wrote a letter to the *Washington Post*, believing they'd take some action. It was her first disappointment in the mainstream media, but certainly not her last.

Luckily, they found a way to get the job done and not put the other lab workers at risk. They decided to work through the night for the most dangerous parts of the procedure, when the pregnant staff were not present. They delivered the purified thirty grams of HTLV-1 and IL-2 on time to the Gallo lab in Bethesda, Maryland.

However, shortly after that, Mark called Judy into his office and handed her a letter. The letter said her job had been eliminated as part of a "reduction in force," ordered by the NCI. This was an initiative pioneered by President Reagan that was a complete scam. Many people would be fired, but it would be called a Reduction in Force (RIF), with the expectation that they'd be rehired as independent contractors, but without the cost of health benefits or pesky rights of employment law.

Later, Judy was told directly that Gallo himself had pulled the strings which led to the firing of the fermentation chemistry team, specifically for their questioning of safety practices.

Aside from people like Gallo, one of the nice things about the NCI was they'd often have lunch time seminars, given by researchers, that anybody could attend. They were also good places to go job-hunting, especially since when you got "fired," there was usually a lag time of months before you had to leave. Judy was fortunate enough to attend a talk given by my friend, Dr. Joost Oppenheim, on the subject on Interleukin-1, a cellular communication protein, which like interferon, was being looked at as a possible cancer treatment. When Judy was twelve years old her beloved grandfather died of lung cancer, and it kindled a desire in her to help cure cancer.

Judy was fascinated by Oppenheim's description of the research, his intelligence and evident goodness, and raced to the podium after he'd finished speaking to talk with him. Oppenheim invited her to have a longer talk with him back at his office, and Judy was excited that she might once again find herself on the leading edge of medicine, developing therapies which might be the cures of the future.

During their discussion, back at his office, Judy let Oppenheim know she was looking for a job and hoped he might have something available in his lab. Oppenheim replied he knew a colleague who'd worked on HTLV-1 and IL-2 who could use a good technician and sent her to me. The interview was scheduled for a few days later and in the interim, Judy read everything she could get her hands on about HTLV-1 and IL-2.

Judy recalls bringing her resume to the meeting and that I was pleasant, but not especially warm. I reviewed her resume and noted with approval that she'd worked in a Biosafety Level 3 facility. Judy mentioned she'd purified IL-2 and HTLV-1, so I naturally asked her, "Oh, have you read the scientific literature concerning HTLV-1 and IL-2?"

"Absolutely," she replied, brimming with confidence.

I was trying to lighten the mood and with a hint of mischief asked, "Do you know who wrote those papers?"

A look of mild panic swept across her face. "Well, no, I don't," she replied cautiously, before her confidence began to reassert itself. "Because it doesn't matter who wrote the papers. I simply read the data so I understand the science and can do the job."

"I wrote them!" Judy recalls me saying, my annoyance plainly visible to her.

I don't quite remember it that way. I may have said those exact words, but I don't remember being annoyed.

However, it was clear to me that Judy was embarrassed, as her face flushed red. She thought she'd completely blown the interview. I saw her eyes move about the room, looking desperately for something to save her. I had a poster on the wall of the Boston Celtics, which at that time included Larry Bird, Robert Parish, and Kevin McHale, arguably the greatest front line in the history of basketball.

In those days, there was a player at the University of Virginia, Judy's alma mater, Ralph Sampson, who was a basketball phenomenon. The Boston Celtics had made an offer to Sampson to leave college at the end of his junior year, but he had not yet replied. "Do you think Ralph Sampson will accept the offer from the Boston Celtics?" she asked.

"He'd be a fool to turn it down," I answered. "If he lets himself go to the draft, he'll be picked up by a weak team that won't be able to protect him from injury. And that will limit his potential." Sampson stood seven feet four inches, but at less than two hundred and thirty pounds, he was quite thin. Although a guy like that can dominate the court, he can also get easily injured. It's simple physics. The bigger you are, the harder you fall.

"I don't think he'll take the offer," Judy replied. "Getting that degree means so much to him and his family." As it turns out, he did not take the offer from the Boston Celtics; he graduated from the University of Virginia in 1984 and was drafted by the Houston Rockets. He was an All-Star for three years, but injuries cut his career short before he could win an NBA championship. Sampson went on to have a successful coaching career at James Madison University and has a lovely family. I still think he would have won multiple NBA championships and had a lovely family.

This simple difference in perspectives, nearly forty years ago, beautifully illustrates so much of my relationship with Judy over the decades. We were both right, with simply a difference of opinion on how we measured success.

Although Judy was certain she'd blown the interview, I was impressed. This was a person who spoke her mind, without worrying much over what I'd think. I went to the human resources manager and asked that she hire Judy.

The human resources manager refused to hire her.

"Why?" I asked.

"Because she's a troublemaker," the human resources manager replied.

"How does she make trouble?"

"She asks too many questions."

"In science you're supposed to ask questions," I shot back. "Hire her or I'll have your job." I didn't usually lose my temper like that, but my experiences with corrupt practices the previous few years had sharpened my tongue. Enough was enough!

On June 6, 1983, Judy Mikovits officially started working for me. We always celebrate that day, and humorously note that it's the same day as the D-Day invasion of France in 1944 during World War II.

Although it took the allies less than a year from the day that they landed on those beaches to defeat the Nazis, our fight for the truth in science has been going on much longer, with the outcome still highly uncertain.

* * *

As a young investigator arriving as an exile from a prominent lab run by a dictator, I probably should have focused my lab on research projects as far away from Gallo as scientifically possible.

However, Dr. Ron Herberman, the program head, constructed a BioSafety Level 3 for me specifically to study these disease-causing human retroviruses. The lab would be in wing three of Building 560 at the Fort Detrick facility, while my clean lab and office were just across the street in Building 567.

One fateful day in the summer of 1983, I was visited by Raoul and Kurt, two senior investigators at the NCI. They had a proposal that we should look for a virus in this new disease called Gay-Related Immune Deficiency (GRID), but which the world would later come to know as Acquired Immune Deficiency Syndrome (AIDS).

Anthony Fauci was then a branch chief at the National Institute for Allergy and Infectious Diseases (NIAID) and he'd been seeing GRID/AIDS patients at the NIH's Clinical Center. Raoul and Kurt talked me into going to a meeting with Fauci in his office. Raoul and Kurt were at the meeting, as well as Cliff, Fauci's clinical point man. I presented our case for a retroviral etiology for GRID/AIDS, and Fauci agreed that NIAID would provide blood samples from suspected patients, subject to two conditions.

The first was that Fauci be given complete authority over authorship on any manuscript, and that Fauci wanted to go on television. (He must love today's exposure.) Cliff talked specifically about wanting to go on *The Tonight Show*, to be interviewed by Johnny Carson.

Alarm bells should have been going off in my head. I do recall thinking, *Right! I discover another virus and I'm sure to get fired again.* I still ask myself to this day why I entered into a collaboration that might once again bring me into conflict with Gallo, and headed as it turned out by a man, Fauci, who apparently had little or no moral compass or integrity.

In May 1983, French researcher Luc Montagnier and his team had reported in the journal, *Science*, their discovery of lymphadenopathy virus (LAV) in GRID/AIDS patients.[4] This virus would eventually be designated the Human-Immunodeficiency Virus (HIV). This paper showed that the virus isolate was not related to HTLV-1. The publication had been largely ignored because Gallo and his associates had three papers published in the same issue of *Science* claiming HTLV-1 *was* linked with GRID/AIDS. Gallo was first author on the major paper (no one stood up to him, unlike with HTLV-1).[5]

The suggestion that only one retrovirus caused every new disease, like AIDS, was at the time, and still is today, profoundly unscientific.

Yet, Gallo was so convincing that almost the entire medical and scientific community believed in the validity of Gallo's HTLV-1 theory as it related to AIDS. The suppositions in those three HTLV-I papers that this was an independent viral isolation was so flimsy they could've been knocked over by the slightest breeze. I found myself wondering what had happened to the critical thinking skills of my colleagues.

Three years earlier, nobody of any importance believed that retroviruses could cause human disease, and now the single disease-causing retrovirus discovered, HTLV-I, was being put forward as the cause of all new disease entities. How quickly they had forgotten that Gallo had always been a serial abuser of the truth.

At least conceptually, science shouldn't be that difficult. But this schizo-phrenic back and forth between retroviruses don't cause *any* problems, to they cause *all* problems, was disturbing to watch play out in real time.

It was painfully obvious to all of us who knew Gallo—save his legion of sycophants—that he was trying to steal the credit by having all retroviruses including Montagnier's LAV (later called HIV) virus to be a member of the family of human retroviruses he discovered. When Gallo was forced to admit that HTLV-I had no connection to HIV or AIDS, he tried to refer to HIV as HTLV-III. It's as if the failure to get the scientific community to believe one lie meant he'd just make up another one. Gallo tried to force a connection where none existed, and his behavior was neither scientifically nor morally correct. His continued inability to follow the science where it was leading was a mark of a mediocre scientist.

To me, Gallo's behavior was a double human tragedy. First, it prevented the world community from recognizing the crucial French discovery of Montagnier as early as possible. Second, it was clear that Gallo had no inter-est in improving public health unless he got the credit.

Gallo had these patients' blood on his hands and few in the scientific community ever called him on it.

* * *

I had met Montagnier twice when he visited Gallo's lab. First, showing him how to grow T-cell cultures and second, how to care for them.

I always found the man to be honest, bright, compassionate, and caring.

I went into the project believing we would either confirm the findings of the Montagnier team or discover additional retroviruses in AIDS patients from different cohorts.

Little did I know that Gallo was already hedging his bets.

The game for me was over before it started.

After a September 1983 AIDS conference in Cold Springs Harbor, Gallo proclaimed that Montagnier did not have a virus because all the HTLV-1 reagents tested on his virus were negative. In other words, the HTLV-1 reagents were the only possible way to test for a retrovirus, and therefore since nothing was detected, nothing existed. I got a message from Montagnier's colleague, Françoise Barré-Sinoussi, through her old mentor at the NIH, telling me that Gallo had requested another sample of Montagnier's virus. Barré-Sinoussi wanted to know my opinion.

My reply was, "Please tell Luc not to do it. Gallo will find some way to steal it."

Unfortunately, Montagnier, whose goal was to help the stricken patients, trusted Gallo and sent a second sample of his virus to the Gallo lab. However, he did make sure the transfer was documented. It turned out that the so-called HTLV-IIIB isolate was genetically identical to one of the isolates sent to him by the French, which is probably why the Nobel Prize committee gave Montagnier and Barré-Sinoussi, and not Gallo, the Nobel Prize in Medicine in 2008.

Meanwhile, as we worked on the isolation of the retrovirus in our lab, it was difficult to maintain cultures because the retrovirus caused the cells to die at a rapid rate. We worked to develop additional human cell lines which would support the growth of these AIDS viruses. First, we received a concentrated blood pack once a week from the NIH blood bank as a source of healthy human lymphocytes. Next, we used our most potent biological preparations of IL-2 to maintain the growth and viability of the infected T-cells. Finally, we used a clone of the human T-cell line, Hut-78, which could support virus production and allow us to grow enough of the virus so we could characterize it.

Our breakthrough was to show we could first infect primary human T-cells with this virus, then showed we could transmit the virus to the Hut-78 cell line.

This clearly showed the LAV virus family was infectious AND transmissible.

We quickly put together a first draft of a manuscript and sent it to our collaborators at NIAID, Anthony Fauci and Cliff. We were excited to get their comments, and since Fauci was in a different institute, we believed he would keep the information confidential.

That was a big mistake and since I was traveling, it put Judy Mikovits in a terrible position.

Unbeknownst to me, Gallo and Fauci were having lunch once a week and swapping information. Research data is supposed to be kept "confidential" until publication, not known outside of the group of scientists working on the problem. How long had this unethical transfer of information been going on? The only explanation which makes sense to me in light of subsequent events, is that Fauci shared our research with Gallo, an ethical violation, in addition to being scientifically corrupt, for together they were making plans to help one another's interests.

I was away at a European conference when Gallo called my lab, with Fauci on the line. Since the manuscript had been sent to Fauci, it was probably thought necessary to have him on the call to give it the appearance of legitimacy.

Judy answered the phone.

Gallo and Fauci identified themselves and said they needed to see a copy of our paper, and, in addition, Judy was to hand over the virus we'd isolated.

Judy said as a technician she didn't have the authority to hand over those materials and they'd have to wait until I returned and take it up with me.

Gallo and Fauci started yelling at Judy that she needed to give them the paper and the virus or else they'd have her fired for insubordination.

Judy reiterated that she didn't have the authority to give it to them.

Gallo and Fauci started yelling even louder and increasing their threats, promising she'd never work anywhere in science.

"Fine! Fire me!" Judy replied and slammed down the phone.

A few lab techs in the outer office heard this conversation and warned her of the power of those two men to harm her career. But Judy wouldn't back down.

Judy was upset she might have ruined her career in science and her dream of going to medical school would be shattered, had several sleepless nights, and a boyfriend even suggested she get a gun for protection.

When I came back to the lab and she told me what had happened, I was shocked. "You did that for me?" I asked.

The tension of the past few weeks must have broken in her because she lashed out, saying, "I didn't do it for you, asshole! I did it because it was the right thing to do. Why is it so difficult for people to have principles?" she asked and stormed out of the lab.

Yes, Judy has a fiery personality.

But, as I learned over the years, it wasn't about personal loyalty to her. It was about having integrity, regardless of whom you're dealing with. In the years to come I'd often delight in telling that story, especially the part where she called me an "asshole."

It never fails to get a laugh out of Judy.

* * *

Shortly after I returned from Europe, I got a call from Gallo. He didn't even make a little small talk but dove right into the reason for his call.

"Frank, what are you doing in Frederick?" he asked.

I replied, "You know damn well what I'm doing considering all the spies you have."

Gallo proceeded to tell me that he, and not Montagnier, had discovered the viral cause of AIDS and had submitted four manuscripts to *Science*. But Gallo was always paranoid about loose ends which might deflect his credit. He said, "The US government does not want to be put in the awkward position of supporting two viruses for the same disease." He wanted me to send my virus to him for comparison.

I said, "You could send me your virus."

The conversation didn't go well after that.

Gallo's boss, Dr. Bruce Chabner, the director of Cancer Treatment at the NCI, the man who canceled my first position at the Frederick facility because of Gallo's pressure, called me up and said I had to send both the manuscript and the virus to Gallo.

I called Gallo and told him, "I'd rather flush the virus and the manuscript down the toilet than give it to you, because I know you'd compromise the truth."

Fortunately, about a month later another researcher, Jay Levy, published the isolation of HIV-1 in Hut-78 cells, confirming Montagnier's work.[6] The next year, our lab published our HIV isolation paper, and Gallo's name wasn't on it.

An associate of Fauci later called me to say, "Tony is sorry about what happened and wants to give you some intramural funds in your lab budget for your lab research."

I understood the catch to accepting the money was that they were trying to bribe me into silence, and Fauci was getting his claws into me for his good friend, Robert Gallo.

"What's the catch?" I replied.

I was told Fauci still wanted control over authorship.

I turned him down.

                                    * * *

I learned one other lesson during this sad affair, which was how far the United States government would go to protect one of its own.

At the US government press conference of April 1984, Secretary of Health and Human Services Margaret Heckler constantly referred to Gallo as "our prominent scientist." Not only did Gallo claim he had discovered the virus, but he had tested it against Montagnier's virus, and they were not the same.[7] I could not believe all the lies which were coming out of Gallo's mouth.

The government jumped in as Gallo's willing partner.

In May 1985, the US government issued a patent for Gallo's AIDS test, yet made no mention of Montagnier's AIDS test patent application filed eighteen months earlier. Did they really overlook Montagnier's patent application? Was it an honest mistake? It's hard to believe that's what it was. There wasn't any larger health crisis at the time than the HIV/AIDS epidemic.

It still bothers me today that the scientific community put up with all of Gallo's machinations, and he escaped without any significant punishment. Had Gallo become like a bank that was too big to fail, no matter how egregious the crimes?

The sad thing is that this mediocre scientist had thrown away all the advantages our discovery of HTLV had given him. In January 1982, a full year ahead of Montagnier obtaining his sample from a French patient, Gallo was sent blood material from ten AIDS patients seen in Los Angeles, California, by a member of the NCI's epidemiology branch. We had already shown Gallo how to isolate and prove a novel retrovirus from T-cells and show that they were causing disease. It was all so straightforward that several labs with virologists would have found the virus. Gallo would have been first. It later turned out all ten samples were positive. Many others published several viral isolates between 1984 and 1985.

What did Gallo do instead of performing the obvious science?

Gallo concocted this crazy theory that HIV was really HTLV-1 and used reagents designed to test for HTLV-1. Then when those showed up negative, he declared there were no retroviruses in the sample. Subsequently, Gallo then claimed to have discovered HTLV-III (which was really HIV and Montagnier's discovery), used different reagents, and found the virus! And all of this was applauded by the sycophants in the US government research community as if he was a brilliant scientist rather than a concocter of fantasies and a thief.

Gallo's approach displayed a complete disregard for the scientific method and squandered the advantages given to him by the work Bernie Poiesz and I had done.

The evidence is clear that Gallo's lab was growing Montagnier's LAV/ HIV virus, and Gallo knew it before his May 1983 publications in *Science*. The viral preparation sent to the Frederick Cancer Research Facility electron microscopy lab by Gallo to visualize the viral particles was labeled LAV, which was the designation given to Montagnier's virus before it was changed to HIV.

An honest mistake by the sender of the sample? Probably not.

One of the French isolates and Gallo's HTLV-IIIB were identical.

As Dr. Gerald Myers wrote in a private letter to Dr. John LaMontagne and other HIV scientists on April 8, 1987, and later quoted in a Congressional investigation:

> [I]t is the astonishing and unforeseen variation of the virus [HIV-1] which exposes the fraud . . . I suggest that was have paid for this deception in more than the usual ways. Scientific fraudulence always costs humanity . . . but here we have been additionally misdirected with regard to the extent of the variation of the virus, which we can ill afford during the dog days of an epidemic let alone during halcyon times.[8]

One of the most disturbing parts of the response of Health and Human Services, as well as the NIH, was how Gallo's interests became the same as those of our government. This meant that these agencies, which were supposed to be objectively protecting scientific integrity, were instead protecting the honor of Robert Gallo from any possible slander. It is shocking to me how quickly public health and science were submerged into defending the behavior of Robert Gallo. I agree with Don Francis, who wrote:

> The facts in the Gallo case are clear. Unbiased scientists in the field know what happened. I, because of my position as head of the CDC's AIDS laboratory, saw it up close. In my opinion, Dr. Gallo's behavior was disgraceful, an insult to the integrity of all scientists. Dr. Gallo purposely tried to rob the credit for the discovery of HIV-1 from the *Institut Pasteur*. It was not passive or an oversight . . . The stories of Dr. Gallo's unprofessional behavior go on and on.[9]

One story which made the rounds was that Gallo was so crazy with paranoia that he told both his gardener and next-door neighbor that he'd been cheated out of not one, but two Nobel prizes.

After my years in government, I learned to be skeptical of any pronouncement from the NIH which might reflect badly on them. But in those days, we had a functioning press, which might occasionally report on government misconduct. As reported on the front page of the *New York Times* on December 31, 1992:

> After three years of investigation, the Federal Office of Research Integrity today found that Dr. Robert C. Gallo, the American co-discoverer of the cause of AIDS, had committed scientific misconduct. The investigators said he had "falsely reported" a critical fact in the scientific paper of 1984 in which he described isolating the virus that causes AIDS.
>
> The new report said Dr. Gallo had intentionally misled colleagues to gain credit for himself and diminish credit due his French competitors. The report also said that his false statement had "impeded potential AIDS research progress" by diverting scientists from potentially fruitful work with French researchers.[10]

In what universe does Gallo escape punishment? But of course, he's still allowed to speak as a leading scientist.

In the wake of Gallo's unholy press conference of April 1983, members of the gay community visited Gallo to discuss how to limit the AIDS epidemic. Gallo threw them out of his office, saying they had caused the epidemic.

Shortly thereafter, the editor of the New York City gay newspaper *Pride* called me, stating they wanted to write a series of articles exposing Gallo's behavior and wanted me to check the science before they published them. I was happy to oblige them. My honorarium for that bit of scientific work was a subscription to *Pride*, which arrived every week in an unmarked brown paper envelope. One of the magazines is still in my possession.

To help give this marginalized group a voice meant more to me than public praise, for through ACT-UP, they changed the scientific and medical response to the AIDS epidemic for the better.

CHAPTER SIX

Failed Public Health Response to HIV/AIDS

Give all the power to the many, they will oppress the few.
Give all the power to the few, they will oppress the many.

—Alexander Hamilton

History is a later event because those involved in the events:
Those who might be able to contradict what is accepted are no more,
Thus, the truth of history is a fable agreed upon.

—Napoleon

By 1985, the US government, Robert Gallo, and Anthony Fauci had gotten what they wanted by messaging scientific truths and throwing competitors under the bus for their own purposes.

The government got a scientific hero who justified all increased funding.

Gallo was recognized as a co-discoverer of HIV, theoretically putting him on the way to winning a Nobel Prize.

Because of his alliance with Gallo, Fauci secured a promotion to the directorship of NIAID in 1984.

There were persistent rumors that the three final candidates for the directorship of NIAID were two members of the national academy and Anthony Fauci. In my estimation, the academy members were both far superior scientists to Fauci, as well as being men of outstanding character. The candidates were told they had to give up their laboratories in order to assume the directorship. The academy members withdrew and Fauci got the position by default and never gave up his lab, proving once again that Fauci's word cannot be trusted.

One of the many terrible decisions made by President Reagan was to eliminate the retirement age for federal employees. As a result, Fauci has been a virtual emperor at NIAID for thirty-seven years, since before many of the current employees were born. The result of Reagan's decision has been that the average age of the men in charge of medical research has substantially increased and women who could be leaders are essentially non-existent. Scientists know that most of the innovative, groundbreaking ideas come from younger researchers in any given field so as NIH becomes older it becomes less relevant.

One of the best things Reagan did, in the face of his drastic funding reductions, including a 25 percent reduction in the Health and Human Services budget, was to give increases for research funding. This led to an expansion of the Biological Response Modifiers Program (BRMP) at the Frederick Cancer Research Facility, directed by Dan Longo from 1985 until its final closure.

However, the program was effectively dismantled in late 1996 when the government declared in its infinite wisdom that the public did not need a translational research component!

The BRMP had basic laboratories to discover useful biological and clinical information, then quickly translate them into therapies. The enthusiasm in the program was so great that an after-work "Amazing Papers Journal Club" sprung up which the researchers eagerly looked forward to each week. They could share their latest findings with all the members of the program, from the most senior scientist to the most recent hire. The BRMP brought together clinicians and basic research scientists focused on understanding the response to therapies designed to modulate the immune system. The biological agents studied by many programs led to a great increase in the understanding of functional cellular immunology, particularly the activation of innate immune response to pathogens, or what we have come to understand better as an inflammatory cytokine storm.

It was the best program Judy and I were ever associated with at the NCI and we never understood why it closed. We were told only that translational research was not needed at the NCI-Frederick.

Sure, we wouldn't want to "cure" cancer at the NCI, would we? What would happen to the jobs of all those researchers?

Our suspicion was that other, more politically connected scientists lobbied better than our advocates for those research dollars. The Society for the Immunotherapy of Cancer later honored our whole team at an awards ceremony at the Smithsonian National Museum of Natural History in 2010 where they said our team had, "imagined, developed and sustained immunotherapy of cancer for more than a quarter century and continued to make seminal contributions to the field."[1]

When Reagan became president, America was the largest creditor in the world, and when he left, we were the largest debtor in the world. I believe his anti-government fervor, deregulation, and massive budget reductions fueled corporate America's exceptional rise in wealth at the expense of the common man.

The "greed is good" approach swept through the health-care industry as it did every other segment of the economy.[2] Twenty-five percent of the federal health-care budget disappeared in a single day, eliminating many worthwhile programs. For example, one program sent physicians into isolated, underserved communities. Reagan's devastating budget cuts resulted in the closure of 250 community health centers and over six hundred rural and urban hospitals, leaving many of these communities without health professionals.[3] Infant mortality increased and African American life expectancy decreased.[4] During the Reagan years, 35 percent of Native American deaths were younger than forty-five years old.

One example of the disastrous results of these decisions was that, in the 1980s, it was claimed that the United States had brought tuberculosis under control. This statement was based on little or no evidence that I could ever discover. As a result, drug resistant TB, which develops when patients do not complete the standard therapy regimen, surged, especially in poorer areas. Because of these budget cuts, Native Americans were 400 percent more likely to die of tuberculosis than other Americans.[5]

I believe Reagan's "pull yourself up by your bootstraps" philosophy also encouraged people to view public health programs as a handout for the poor. The legacy of medical misogyny and racism has a long history in our country, and these devastating cuts made the situation even worse. Minority groups and people of different sexual orientations are generally those most damaged by the system and more reluctant to seek treatment for problems. Unfortunately, it is among the conditions of poverty that most diseases arise.

Before COVID-19 hit in 2020 many people were dying in America, with 70 percent of these deaths linked to chronic diseases or medical "mistakes," such as the wrong medication or dosage. In 2018–2019, it was estimated that these medical "mistakes" were the third leading cause of death in our country with, depending on the study, between 250,000 and 440,000 victims.[6]

Several studies have shown a curious pattern that American citizens in the twenties and thirties do not have the same robust health that their parents did.[7] Who is in control of regulating air quality, food safety, water purity, and the very medications we put into our bodies? All these things have become more or less privatized, legal protections have been loosened, and companies have been given "immunity" from being sued, and I credit much of this to Reagan's mania for deregulation. It seems like-minded Democrats have no interest in fixing the system either, I have reluctantly concluded. Consumers have no protection from the harm inflicted by unaccountable corporations, and as a result, we are getting sicker and sicker.

Ironically, as an editor of the *New England Journal of Medicine* remarked in 1986, a medical industrial complex was developing because of these actions.[8] Reagan's policies allowed drug makers, dialysis companies, and others providing profitable, high-cost medical treatments to greatly expand their influence.

In my estimation, it all began with the Bayh-Dole Act of 1980 signed into law by President Jimmy Carter. This act allowed universities, small businesses, and other non-profit organizations to elect to retain any patentable invention made with federal funds. That's right. The taxpayers funded it, but it's owned by somebody else. The institution retaining the title must commit to commercialization of the invention. The university is also required to share a portion of the royalties with the inventors and must also dedicate a portion of the money earned to laboratory purposes, such as buying new equipment and performing maintenance.

In the forty years since the act's passage, Big Pharma and the universities have not become centers of innovation, as the act was intended to do. In fact, none of the universities or Big Pharma companies which received money because of Bayh-Dole created their own drugs or therapies.[9] A study reviewing the 210 medicines approved by the FDA from 2010 to 2016 found that all of them had been developed from NIH-funded research.[10] In other words, the money from the inventions went somewhere other than to the taxpayer. The money went to Big Pharma and the stockholders, not the US taxpayers, the patients and stakeholders.[11] In my estimation, universities aren't using the money as a public trust to fund the next generation of therapies. Instead, Big Pharma is probably just lining the pockets of its executives or Wall Street profiteers.

For example, the NIH funded sixty-two million dollars for the development of a Hepatitis C drug to a company called Pharmasset. That company was then bought for eleven billion dollars by another company, Gilead, which then decided to price the treatment at six figures, even though the treatment itself costs less than a hundred dollars to produce. From a *New York Times* article on the sale:

> Gilead Science made a bold move on Monday to capture the lead in developing the next generation of hepatitis C drugs, agreeing to pay $11 billion in cash for Pharmasset.
>
> The treatment of hepatitis C has undergone a revolution this year, with new pills from Vertex Pharmaceuticals and Merck sharply increasing the cure rates and also often cutting the required duration of treatment. But those new drugs still must be used with alpha interferon, a type of drug injected once a week that can cause severe flulike symptoms and other side effects.
>
> Pharmasset, based in Princeton, N.J., is pushing to develop the first all-oral treatment regimen, doing away with the need for interferon.[12]

Stories like that turn my stomach. Why are scientists interested in becoming billionaires? This is not science, it's commerce.

This is what our tax dollars are being spent on.

Seed money goes to companies, then they make billions by jacking up the price of their treatment.

Is it clear why so many scientists, like Harold Varmus, travel so easily between government work and outside groups? The system works great for insiders who want to cash in. Maybe somewhere in their souls they believe they're helping people, but I doubt it.

Furthermore, by allowing universities and Big Pharma to have patent monopolies, Bayh-Dole inhibited the open flow of knowledge so other scientists could not use the information to develop better and safer therapies, or ones that cost less.

Do you really believe that these organizations will kill the goose that lays so many golden eggs for them?

For many common diseases, progress in treating them has slowed to a crawl. The laborious construction and development of knowledge paid for by the taxpayer must be communicated in order to fulfill the scholar's obligation.

Another example is that since the development of recombinant insulin in 1977, age-related deaths due to diabetes have increased. The $250 billion industry for the treatment of diabetes has inhibited a public health campaign to greatly decrease the incidence of diabetes through dietary changes.[13] This slow-down in innovation has had a harmful effect on urgent research needs in essentially all chronic and infectious diseases.

Since only about 5 percent of drugs designed to treat neurodegenerative diseases make their way to clinical trials, Big Pharma has essentially stopped all research into these diseases.[14] Second, there exists a drought in alternative strategies of antibiotic drug development. Drug companies do not do this research because antibiotics are not very profitable, in that they are only needed for a short time and the pathogens become resistant to them. Those which do remain in use eventually go off-patent and are inexpensive, thus not greatly contributing to the company's bottom line. Other technologies, like specific bacteriophages which would be used to kill drug resistant bacteria, are not being pursued with any vigor.

A publicly funded system of drug development in this arena is urgently needed and would be less costly to the taxpayers.

The worst result has been the corruption of evidence-based medicine, which results in the all-too-often killing of the patient for profit. In 2002, Dr. Relman, former editor-in-chief of the *New England Journal of Medicine*, said,

> The medical profession is being bought by the pharmaceutical industry, not only in terms of the practice of medicine, but also in terms of teaching and research. The academic institutions of this country are allowing themselves to be paid agents of the pharmaceutical industry. I think it's disgraceful.[15]

Dr. Relman's replacement as editor-in-chief at the prestigious publication sounded a similar warning note in 2009: "It is simply no longer possible to believe much of the clinical research that is published, or to rely on the judgment of trusted physicians."[16]

It is not possible to discuss all the reasons for this sorry state of affairs, but I can briefly mention some of the most glaring examples. They would include: the rigging of outcomes of clinical trials, selective and biased publications, commercials for drugs posing as publications written by industry scientists (learned from the tobacco industry), and the outright bribery of journal editors. As has been said by many others, some universities and doctors have become willing participants in this game of killing for profit.

Never has this corruption been more apparent to me than during the COVID-19 crisis.

But surely, we can expect our huge number of regulatory bodies to protect us, like the Centers for Disease Control and Prevention (CDC) or Food and Drug Administration (FDA)?

We cannot. Since Reagan's time these agencies have been consistently weakened to the point where they have become inept and useless. In my estimation, Republicans do the most damage to kill these agencies, but by the same token, when Democrats come to power, they do little to restore them.

For me it's easy to identify at least three "man-made plagues" caused by this lack of effective oversight.

First, many FDA failures have contributed to the opioid epidemic. Opioid use disorder is now common in patients, and opioid overdose is the leading cause of accidental death.[17] Proper oversight could have prevented most of this tragedy, as could the federal legalization of cannabis and FDA approval of Peptide T and low dose interferon alpha.

Second, after lawsuits against vaccine manufacturers and health care providers threatened to cause vaccine shortages, the National Childhood Vaccine Injury Act of 1986 was passed, spearheaded by Democratic Congressman Henry Waxman of California and reluctantly signed into law by President Reagan.[18] This act removed all legal liability from vaccine makers and instead put the federal government on the hook for any vaccine injuries, creating in effect a special "Vaccine Court" with its own rules of evidence and capping damages. This was tantamount to letting the fox guard the henhouse.

With the Department of Justice and its battalion of lawyers deployed against the distraught and exhausted parents of vaccine-injured children in what was claimed to be a "no-fault system" of compensation, it was clear that no incentive existed for vaccine makers to improve the safety of their products.

A counterpoint to this act was occurring at the time because the "blood shield laws" financially protected the blood banks for giving HIV to patients through a contaminated blood supply.

Third, in what I consider one of the most cynical and tragic of the man-made plagues, I talk about the so-called "War on Drugs." Even from its inception, it was a scam. For decades, the federal government has secretly engaged in a shifting series of alliances with some of the most notorious drug cartels to ever exist, often providing them with the highest level of protection, usually through the CIA.[19] This included cartels run by the Mexicans, Panamanians, and Columbians. American drug agents were allowed to be captured and killed by these cartels, in what I consider to be the ultimate act of betrayal.

While Nancy Reagan was imploring inner-city youth to "just say no" to drugs, her husband was running a secret war in Nicaragua, financed by the Contras who were selling cocaine in American cities. This resulted in a five- to ten-fold increase in the number of people jailed for drug offenses.[20]

The government loves to tell you it's doing one thing, like protecting you from illegal drugs, or COVID-19, when their real agenda is starkly different.

The criminalization of cannabis is, in my mind, a great crime, especially as it created a near black-out of research on the possible medical uses of cannabis. Even in 2017, the attorney general called for a crackdown on marijuana usage.

In government, it just seems you can never quite kill a bad idea. Proper regulatory oversight could have prevented these needless tragedies.

Is it any wonder that the Reagan administration's response to the AIDS epidemic was to stick its head in the sand and not see anything from 1983 to 1987? In 1985, the administration rejected an AIDS prevention plan, telling the public health officials to basically "look pretty and do as little as you can." As a result, by the end of 1987, sixteen thousand blood transfusion recipients and ten thousand hemophiliacs (who required Factor VIII, a clotting product made from human blood) became infected with HIV.[21]

Meanwhile, we were studying the effects of these retroviruses on the overproduction of immune factors in the body, which were likely involved in the pathogenesis of AIDS. Due to the extremely fragile nature of these retroviruses, the isolation of them for study can be extremely difficult. Judy was excellent in developing a two-step hollow fiber tangential flow filtration process for retroviral purification. First, a microfiltration step removed cells and large debris, and then an ultrafiltration step concentrated viral particles. We then used this technology to publish the isolation of HIV from an ALS patient in 1985.[22] Judy's creative engineering solution to this problem was critical to our success.

It was not the only area in which she made her presence known.

* * *

Working on the Fort Detrick military base as a civilian employee certainly has its share of ups and downs.

Civilians like Judy and I could use most of the facilities at the base, like the excellent gym and community center. There was also a squash court where Judy taught me a variety of that great game, using a slow ball, which is more physically demanding and designed for more advanced players. On the rare occasion that I came dangerously close to winning, I would often be hit in my buttocks with the squash ball as I had taken the wrong position on the court. That is, I dared to take the command center position. She let me know in no uncertain terms what could happen if I wasn't prepared on HER court.

At least, that was usually her explanation as to why I'd been hit by the squash ball at the most sensitive place on my upper back thigh.

The painful contusion often lasted weeks, and I was careful the next time to make certain I really had command of the court before I took that position. I knew my shot had to be precise enough to be impossible for Judy to get a racket on.

Judy also played ice hockey, and as in soccer and squash, she often played the defensive goon as if the fate of Western civilization rested on the outcome. No one was coming into her territory without paying the price. We coined our squash game "New York Islanders squash" after the professional hockey team. As when we played basketball, no referee was needed.

It was simple. No blood? No foul!

When she'd go jogging with the young, male seminar speakers, they'd often return complaining of bad knees or aching hamstring muscles as the reason they couldn't keep up with her fast pace.

She organized and coached an annual softball team, relishing her position as catcher where she'd taunt the big, strong military men as to why they couldn't hit an easy pitch over the fence. They'd usually respond by swinging so hard that they'd hit a simple pop-fly, for which we'd immediately call out "can of corn!" In one game we called "can of corn" ten times and one of the players challenged us as to whether that was a real expression in baseball or something we'd made up.

After the game, we went to a baseball almanac to try and find the answer. Apparently, the term came from grocery clerks who'd knock a can off a high shelf with a stick, then catch the falling item in their aprons.

One time the other team was getting so rattled by Judy's constant stream of talking during a championship game that the umpire told Judy if she said one more word, he'd throw her out of the game. She complied, and we won the game.

I kept the trophy in my office until I retired, not because I was proud of the team, but because it was demonstrable, undeniable scientific evidence that it was physically possible for Judy to occasionally keep her mouth shut.

It was never boring to work with that woman.

Like me, she never had much patience for the constant useless challenges by some of the military authorities at the base.

Judy rode her bike to work almost every day, always before dawn, when the sleepy Frederick, Maryland traffic was non-existent. One day she forgot her helmet and the guard at the gate told her to ride home and get it. Helmets were mandatory when riding on the base.

"It's five miles back to my place," Judy said, "and there's a lot of traffic. Wouldn't it be safer to just give me a warning and I'll remember it tomorrow?"

The guard told her she needed to go home.

Judy ignored him and quickly rode onto the base, hoping he didn't know where she worked.

Later that morning, military police dragged Judy out of the laboratory and forced her to ride home without a helmet. It reminded me of how the military police used to come for me and march me between two guards so everybody could see, just to escort me to the barber shop because my hair was too long.

When it comes to mindless authority, Judy and I really are two of a kind!

* * *

In 1985, the US military created an HIV research program which was funded mainly by Anthony Fauci's NIAID.

However, when I tried to talk to some of the scientists about their efforts, I was told I had to get permission from the military to talk to any of the scientists. Can anybody tell me why the military was interested in concealing the research they were doing on a disease which mainly affected gay men? Remember, according to the military at that time, there were no gay men serving in uniform.

While I was interested in doing basic research, and not therapies, that all changed when I met a brilliant researcher, Dr. Candace Pert. Like Judy, I felt she was a kindred spirit. Her discovery of endorphins (chemicals which provide a natural high, such as after exercise) while a graduate student was made because of an experiment she conducted, in spite of her graduate advisor telling her not to do it.

When her advisor, Solomon Snyder, was awarded the Lasker Prize for her work, without any attribution to her, she would not stay quiet.[23] She became a pariah in the misogynistic patriarchal scientific community, just as Judy would become years later. When Candace reported that both insulin and endorphins were made in immune cells in the brain, her papers were roundly rejected by the scientific journals with comments like "are you washing your test tubes enough?"[24]

After her work was later confirmed, the journal *Nature* warned scientists to beware of those "radical psycho-immunologists" who would suggest that the mind and body are in communication.[25] Candace Pert would later be nicknamed the "Mother of Psycho-neuroimmunology" and the "Goddess of Neuroscience."

This connection between the neurological and immune systems was enough for Candace and her immunologist husband, Mike Ruff, to make a peptide that would mimic the part of the virus which binds to the receptor on the cell surface which allows the virus to gain entry to the cell. Candace asked Judy and I to do the HIV infection experiments with the peptide they'd dubbed "Peptide T," which was an eight amino acid peptide within an HIV envelope, the outside portion of the virus containing the gp120 spike protein of HIV. We showed that it would block gp120 from binding to the CD4 molecule on helper T-cells and that the CD4 molecule was present in the brain, probably on microglia, not T-cells.[26]

Beginning in 1986, we published a series of papers in peer-reviewed journals stating that Peptide T inhibited the infection of HIV viruses that used the CCR5 receptor to enter immune cells. It was not known that HIV used the CCR5 receptor to enter immune cells until 1995.[27] Since Peptide T does not completely inhibit the isolation of all viral isolates, this work was roundly dismissed by the mainstream medical AIDS community. It was particularly harsh because the NIH was completely invested in pushing azidothymidine (AZT), despite its profound toxicity. Candace said many times that we did not expect Peptide T to be a cure, but it could be part of a cure. Even when we showed that Peptide T blocked HIV binding to the CCR5 receptor and did not inhibit the lab-adapted viruses which bind to the CXCR4 receptor as a cofactor, nobody was interested in Peptide T as a drug.

In 2007, twenty-one years later, an independent team confirmed our results.[28] They even extended our findings to suggest that Peptide T, alone or in combination with other anti-retrovirals, could prevent infection of the central nervous system and the resulting neuronal damage that leads to neuro-AIDS.

Much of this story is memorialized in the Oscar-winning film, *The Dallas Buyer's Club*, starring Matthew McConaughey.

* * *

A scholar's obligation is not only to generate knowledge, but to communicate and defend the data at all costs. That is why our 2015 consulting company mission statement included the Thomas Jefferson quote, "For here we are not afraid to follow the truth wherever it may lead; nor to tolerate error so long as reason is left free to combat it."

Judy likes to say, "God has a sense of humor."

I think about it differently. Scientific advancement is a mysteriously beautiful thing, as the truth may be hidden for a time but can often reappear in a different form. RAP-103 is a stabilized analog of Peptide T and is being used as an inhibitor of neuropathic pain. That means its safety has been established and, given a few simple government permissions, we could use it today to start treating people for various conditions, including HIV/AIDS and SARS/AIDS, termed "Long-haul COVID."

How else might RAP-103 (Peptide T analog) be used? In 2019, it was shown in mice that CCR5 levels skyrocket after a stroke, and it is this inflammation which causes a great deal of the damage.[29] But RAP-103 (Peptide T analog) targets the CCR5 receptors, meaning it could limit the damage caused by a stroke. Isn't it interesting that COVID inoculations have strokes and blood clots as serious side effects?

Exactly what the data would suggest would happen when expressing viral envelopes.

Another drug, Maraviroc, also targets these receptors, and was approved in 2007 as an antiviral in HIV infection. (Isn't it curious how many of these compounds which target the immune system are also good against viruses?) However, Maraviroc also has numerous documented side effects, including liver damage and allergic reactions.[30]

If given a chance, wouldn't you rather try Peptide T or RAP-103, which have been found to have little or no side effects, rather than Maraviroc?

But under current FDA pronouncements, you will never be able to make that decision.

We can only hope that someone tries to genuinely revive Peptide T as a therapy sometime in the future. We believe an immediate use for Peptide T would be for people who suffered a stroke after receiving their COVID-19 vaccine.

A final footnote to the Candace Pert story is that in 1999 she autographed my copy of her book, *Molecules of Emotion*. I always thought it read, "To a man with fatal integrity."

Judy looked and it and claimed it read, "To a man with total integrity."

We laughed about it because both claims were equally plausible. In an era of corruption in so many professions, total integrity can be fatal to your career.

Right up until her untimely death in 2013, Dr. Candace Pert was a true scholar in this age of corruption. More importantly, she did it with aplomb and humor. When Judy gets discouraged by the attacks surrounding her discoveries about XMRVs and chronic fatigue syndrome, she often recites the quote that was on the poster in Candace's office. It read, "When they are chasing you out of town, simply run out in front and pretend you're leading the parade!" This quote was also included in Candace's obituary, and is a fitting tribute to a great scientist.

It matters little what people say about Judy or me. The only thing that matters is having total integrity in the interpretation of the scientific data, no matter how inconvenient those truths may be to those in positions of power.

In the summer of 1986, I thought my collaboration with Judy Mikovits would be permanently ended when she took a position with Upjohn Pharmaceuticals in Kalamazoo, Michigan. As a woman stuck in government science, which was run by tyrannical, women-hating men, she saw little chance for advancement.

But before she left, she gave me a parting gift.

I had never learned to drive, and she harangued me, gave me driving lessons, and made sure I got my license.

The job in Kalamazoo didn't work out, as she ran afoul of tyrannical men there as well, and I welcomed her back, helping her get into a doctoral program at George Washington University.

CHAPTER SEVEN

Political and Research
Struggles Continue

*The true enemy of science is the substitution of thought,
reflection, and curiosity with dogma.*

—Frans De Waal[1]

*The difficulty lies not so much in developing new ideas as in escaping
from the old ones.*

—John Maynard Keynes[2]

Our son, the greatest of Sandy's and my experiments, was born in
Washington, DC in 1986.

It was a difficult delivery. After sixteen hours I was ordered out of the
delivery room, just as I heard a loudspeaker announcing "Stat!" An eternity
of an hour later, I was holding this tiny person in my arms. Both of us were
confused by what happened next.

All seemed well, except that Sandy had not seen our new son. He had a
bacterial infection (perhaps a result of being unattended in the birth canal
for so long), and we had to consent to a spinal tap to judge the extent of the
infection. But it all ended well, and within a few days he was given the "all
clear" and our small family headed home.

While family life was headed in a delightful new direction, there was
much confusion in my scientific career. It would be many years before I
would be invited to an HTLV-1 scientific meeting. Ironically, scientists

would ask me for invitations to HTLV-i meetings, to which I wasn't invited. My exile was so complete that for thirty years I never received an application from a fellow interested in studying human virology. I can only assume this was due to Gallo's de facto *fatwa* against me.

In the short term, I decided to study the mechanisms by which cells communicate, primarily through messenger molecules called *cytokines*, which simply means soluble factors promoting communication between cells. The yin-yang model that most human systems consist of complimentary parts, one having positive effects, and the other negative, was having an increasing impact on my thinking. We became concerned because, as we tried to determine clinically the maximally tolerated dose of IL-2, we saw that the high doses usually resulted in such toxicity that many patients died. We decided to examine cytokines for their ability to stop an event (i.e., cell growth, immune reactivity, inflammation) in the presence of factors which would stimulate these events.

We searched for a molecule that was available in purified form, and for reagents, which could specifically block the activity of this molecule. A chance meeting with Larry Ellingsworth and Jim Dasch of Collagen Corporation introduced me to the reagents available for Transforming Growth Factor-Beta (TGF-beta), which turned out to be an essential regulator for most, if not all the body's organ systems.

The late Anita Roberts, a wonderful scientist and even better human being, always said that if a cell was off, beta would turn it on.

For the next twenty years, we and many others found that TGF-beta had profound effects on every stage of hematopoiesis, from the stem cell to reactivity of the immune system.[3] TGF-beta antibody Jim Dasch developed turned out to be a very valuable reagent. Jim gave the cells producing the antibody to me and the American Tissue Culture Collection (ATCC) so that all scientists could have access to the antibodies. As usual, no good deed goes unpunished, and years later I got into trouble with the company for freely giving the reagents to other investigators.

In the interim, studying the complexity of TGF-beta was very rewarding and substantially added to our knowledge.

Meanwhile, all through the eighties, there was, generously stated, a lukewarm response to the HIV/AIDS epidemic. It was becoming more and more apparent that the public health establishment was failing to provide the required leadership. The heads of the CDC, NIAID, and the secretary of Health and Human Services (HHS) did not sufficiently share the scientific concerns about the safety of the blood supply and blood products.

This laisse-faire approach greatly impacted the health of three groups: blood donors, hemophiliacs, and the gay community. The most troubling aspect of this epidemic was the number of people fatally infected because of the failures of our health-care system itself. Public health officials in other countries either lost their positions or served jail time for these failures.

But not in the good ole' USA.

In 1995, the Institute of Medicine (IOM) produced an exhaustive 350-page study, "HIV and the Blood Supply: An Analysis of Crisis Decision Making," which squarely placed the responsibility for the spread of AIDS through blood transfusion and blood derived products on the FDA, blood banks, and the blood products industry.[4]

In 1983–1984, nearly two years after several committee meetings discussed the CDC recommendation of blood testing and donor screening, the FDA was still stalling and blood industry spokesmen were still denying that AIDS was a threat to the blood supply. Privately, these administrators were saying something different about the danger, but were content waiting on an HIV blood test when several procedures were available to save lives.

Finally, on March 2, 1985, HHS Secretary Margaret Heckler announced the first AIDS test, which measured antibodies to HIV, and was licensed to Abbott Lab.[5] By July 1985, the FDA recommended, BUT DID NOT REQUIRE, the use of the test, which resulted in many blood banks not testing their blood supply.

To nobody's surprise, this was later called "an honest mistake."

The cowardly, incompetent FDA relied on its friends in the blood bank industry to both advise on the decision-making and to do voluntary testing of the blood supply. Mandatory testing did not become law until February 4, 1988, nearly three years later, after thousands had become unnecessarily infected and died as a result. It is hard to know whether the FDA or the blood banks committed a greater crime against humanity.

This government failure is especially tragic since the blood bank industry is immune from tort liability through shield laws, enacted by Congress. Thus, government regulation is the only currently existing mechanism for consumer protection. The IOM indicted the FDA for relying solely on the industry it regulates for advice. This should have been sufficient warning for why it was a bad idea to grant the pharmaceutical companies' complete immunity for harm caused by childhood vaccines through the 1986 National Childhood Vaccine Injury Act.

Yes, that's right.

The regulators (NIAID, CDC, and FDA) in charge of protecting the consumer against vaccine abuses again left that responsibility to the industry foxes in the chicken coop.

All these same organizations were asleep at the switch with fifteen years of warnings of the dangers of a coronavirus epidemic. No test or therapeutic studies were funded in preparation. Instead, as always, these serial felons simply waited for the pandemic to strike down millions, then waited for the profit-guided vaccine industry to save us all. How long will this nightmare of unethical conduct of public health be allowed to continue? The world has been waiting almost forty years for that miracle HIV vaccine to save us all. One must fund the development of therapeutics to use at the beginning of a pandemic and before if we have had warnings.

* * *

For hemophiliacs, corporate greed and regulatory incompetence provided a double whammy.

These desperately ill patients needed a medication which allowed their blood to clot, called anti-hemophiliac factor (AHF). Since AHF was produced from pooled plasma from thousands of blood donors, it was likely that AHF was contaminated with blood borne pathogens. The assumptions in the hemophiliac community about the medically acceptable risks of this blood product transmitting viruses, such as hepatitis, led to a general failure to react to reports of a new infection among large groups of people.

Harvey Alter, now a Nobel Prize laureate, who we were to meet later, along with many other hepatitis doctors, stated of the problem, "We were used to dealing with hepatitis. You get it from blood. No one got sick right away . . . and it was not life threatening."[6]

Hepatitis was an acceptable risk.

And HIV/AIDS was put in the same category.

How could they do this?

Had they never looked under a microscope, as I had done, and seen the rapid lysis of T-cells from AIDS, unlike any other viral disease in human history? The IOM report found that scientific uncertainty allowed a pattern of responses that were very cautious and exposed the decision makers and their organizations to a minimum of criticism, if they were later found to be wrong.

I understand that in any crisis, mistakes will be made and opportunities for appropriate response will be squandered. However, where there are structural problems in the decision-making process, they must be pointed out. The decision-making committees of public health had members representing blood banks and the manufacturers of blood products, but no opposing input from consumers, independent scientists, or even from the FDA. In fact, the FDA was letting the industry self-regulate itself.

There was enough time to correct any of these decisions, but the accumulating evidence was not critically analyzed. From 1983–1985, there is no evidence that any of these decisions were being made by the top leadership. One of the difficulties with using "experts" to give advice to government agencies is that these experts develop close relationships and accumulate favors with experts on regulatory agencies and the industry itself.

The most astonishing FDA decision was not to demand an automatic recall of AHF products linked to an infected donor.[7] So, throughout the 1980s and early 1990s more than ten thousand US hemophiliac patients became infected with HIV.

In March 1983, Baxter Healthcare was granted a license for heat-inactivated AHF, but Bayer/Cutter continued using the unsafe AHF, while bad mouthing the heat-inactivated AHF. In October 1984, a Cutter study found that their heat inactivated product killed HIV, while 75 percent of the patients who used the unheated products eventually tested positive for HIV.

In November 1984, Cutter told overseas markets in Asia, Latin America, and Europe that "we must use up stocks of unheated AHF before switching to safer heat-treated blood products."[8] Why? Was it done for purely monetary gains? Of course!

In May 1985, when the FDA realized that companies were still selling unsafe, non-heat-treated materials, they told the companies they wanted it stopped, but did not alert the public.[9]

However, the damage was done, and people continued to get infected with contaminated AHF around the world for a decade.

* * *

Meanwhile, Judy Mikovits, working for Upjohn Pharmaceuticals at this time, and a few of the lead scientists, Janie and Steve, saw an urgent need to test the Upjohn biological products made from human blood.

One such product was called ATGAM, human and thymus globulin made by injecting human T-cells (the same cell targeted by HIV) into horses and purifying the T-cell antibodies. ATGAM is still used today to treat some patients with severe aplastic anemia, a potentially deadly blood disorder, where several cell types in the peripheral blood are poorly produced.

State officials in Michigan had proclaimed there was no HIV virus or AIDS in the entire state. This was a ludicrous statement at the time, as Judy had to go no further than her hairdresser in Michigan to find somebody who had AIDS.

Had she inadvertently stumbled onto the single gay person in Michigan with AIDS?

Judy's proposal to Upjohn on the ATGAM issue was to spike a batch of ATGAM with purified HIV, then test it as it went through the manufacturing process to demonstrate a six-log reduction of detectable virus.

Judy and I tested the product after various stages of the manufacturing process, then at the end. We added some steps to the process to ensure a virus-free product and Upjohn accepted the proposal. Our research showed the product was safe under their manufacturing process, but we added some extra steps as a precaution.

Since there was no HIV in Michigan (wink, wink), it was agreed that Judy would conduct the experiments in my Biosafety Level 3 Lab at the NCI. Judy would fly out on Monday morning, using the Upjohn private jet, on which top executives would fly for their weekly meetings with the FDA. It was all very hush, hush, as nobody wanted to say the word *HIV*. The Upjohn officials, all dressed up in their three-piece suits, would stare in confusion at this twenty-eight-year-old blonde woman in tennis shoes and jeans, carrying a backpack, who would chat amiably with them about sports or other issues in science, but reveal nothing about what she was doing. When the plane arrived at National Airport (now Reagan National), a rental car would be waiting for her on the tarmac and she'd speed off to Fort Detrick on her mysterious mission.

So, it was very clear to us the FDA, NIAID, and CDC were at fault and knew exactly what was happening but would simply say that this is mere hindsight. To those of us doing the science and not prone to making unfounded, profit-oriented assumptions, it was not hindsight.

The IOM report of the failure with the HIV-contaminated AHF drug found that there were several risk reduction strategies which were not pursued, or even recommended. As the IOM report stated, "the perfect should not be the enemy of the good."[10]

Obviously, Dr. Fauci, who was a significant part of the AIDS failures, simply forgot to implement these risk reduction strategies in his latest pandemic with his attack on the Henry Ford Health study, which showed that hydroxychloroquine and corticosteroids were effective in lowering the COVID-19 death rate, as flawed. In direct opposition to the IOM report, Fauci said health officials should rely on the gold standard of a randomized, placebo-controlled study. The results with interferon type I and III and ivermectin were also ignored.

How strange it is then, that in 2015, Dr. Fauci and the FDA ignored a double-blind, placebo-controlled study showing that Suramin, a medication in use for more than a hundred years, designated by the World Health Organization (WHO) as an "essential medicine" and manufactured by Bayer/Monsanto, could be effective as a treatment for autism. We have known since 1980 that Suramin was not only a potent inhibitor of the reverse transcriptase enzyme, but also a potent inhibitor of the mouse (murine) leukemia viruses.

Hydroxychloroquine (HCQ) and steroids, along with interferon alpha, ivermectin, and antibody cocktails, may not have prevented the development of COVID-19 in millions of people, but would have prevented the spread of SARS-CoV-2 and greatly reduced the suffering and recovery time of those afflicted. You might be surprised to know that in 2015, Dr. Fauci even went so far as to call HCQ a "vaccine" against the original SARS virus.

In addition, the CDC rolled out a contaminated unvalidated COVID-19 PCR test as the new Gold Standard for diagnosis of an infectious disease, which is what happens when we do not allow others to make competing products.

* * *

During this pandemic, it was and is not only unethical but criminal to hide behind perfect clinical trials, particularly since the decision makers' friends in the pharmaceutical industry have been manipulating clinical trial endpoints for the success of their own products' trials for decades.[11]

The concept that agencies like the FDA should not rely upon the entities they regulate for the analysis of the data led to further questions in the search for AIDS drugs. Under enormous pressure from patients, the FDA fast-tracked the review of azidothymidine (AZT). The toxic effects of AZT on bone marrow suppression, anemia, and T-cell death were already known and observed during clinical trials. Yet, this drug was still deemed safe and the FDA pushed for approval, while safer and more efficacious medications, like interferon alpha and Peptide T, not only did not receive

FDA approval, but their use was forbidden or made illegal. Again throw out the good searching for the perfect.

Not unexpectedly, severe toxicity from AZT resulted in 50 percent of AIDS patients having to be taken off the drug, with many of them dying.[12] Despite some early benefits for AIDS patients, AZT was not found to benefit patients enough. In many instances, the virus had mutated to resist the drug.

Despite this, Dr. Fauci, who had published results that in some cases AZT did not block HIV replication *in vitro*, still pushed for prescribing AZT to asymptomatic people.[13] This caused many AIDS doctors to be outspokenly dumbfounded. Once again, with COVID-19, Dr. Fauci and his fellow criminals in the FDA have used the same playbook with hydroxychloroquine (HCQ), ivermectin, and remdesevir, when remedies like HCQ and ivermectin with much better safety profiles were ignored or made illegal.

The high cost of AZT (at the time the costliest drug on the market), the lack of health insurance for infected individuals, the poor therapeutic window, slow pace for drug discovery and implementation, and social stigma attached to AIDS led to the rise of advocacy groups like ACT-UP, which I mentioned in an earlier chapter.

The important parts of the ACT-UP agenda included: that people from all affected populations at all stages of HIV infection be included in clinical trials or use parallel tracks; that no more double-blind placebo trials be necessary in fatal diseases because they are unethical; that drugs would become available when proven safe (as was the FDA's only mission from its inception); and finally, that drug therapies must be paid for by private or public medical insurance.

Progress was slow, with officials stonewalling protestors until they invaded the NIH in May 1990. With shouts of "Fuck Fauci!" rising from the crowd, changes were being forced on the medical community by the patients for the first time. If only those changes had not been ignored over the past three decades as all liability was removed from the manufacturers of vaccines, but also from harm caused by the nation's unsafe blood supply and biological therapies, untold suffering and millions of lives could have been saved.

In 1990, by the CDC's own definition, women did not get infected with HIV, they just died of AIDS.[14] One of the least known of ACT-UP's priorities was the demand that the definition of AIDS be changed so that HIV-infected women could get treatment and support.

In 2020, forty years since the beginning of the HIV/AIDS epidemic, AIDS remains the leading cause of death of women of reproductive age around the world.[15] This tragedy is the result of gender prejudice, sexual violence, educational discrimination, and lack of reproductive health services. Why is so little being done to recognize the rights of women?

The public believes that government health agencies are thinking and planning far into the future to deal with pathogenic threats to the citizens. Good luck with that fantasy.

The improvements suggested in the IOM report on how the public health agencies dealt with the threat to the blood supply from HIV have only received lip service. Instead, it's the same "old boys club," where corporate greed is an important criterion in all decision making and where unethical if not criminal conduct occurs with impunity.

* * *

In August 1987, Judy's beloved stepfather Ken was diagnosed with stage four prostate cancer.

Ken was in his early fifties and did not want his children to know, as they'd lost their mother to aggressive breast cancer when she was in her thirties. As Judy and Kent wrote in *Plague of Corruption*, she was in a wee bit of trouble at Upjohn for throwing a notebook at her boss, Frisbee style, when he told her that she was "legally, ethically, and morally" required to do what he said, even if meant she would lie and misrepresent data. Here's a hint to anybody who might decide in the future to employ Judy.

Never ask her to lie.

It won't end well.

Judy came back to the East Coast to help her mom and stepfather. I funded her work at the NCI and there was also the possibility that some cutting-edge research might hold a miracle for her stepfather. Judy also decided to begin graduate school, disgusted by what she'd experienced at Upjohn. I knew it wasn't and would not be easy for her to overcome the demons of those degrading misogynistic experiences at UVA, but I also knew once Judy made a promise, nothing would stop her from doing all in her power to achieve the goal.

Thus, the proverbial bad penny returned to my life, bringing her insane work ethic with her. Normally, she'd arrive in the lab at 4 or 5 a.m. If I arrived earlier than her, she'd be irritated and come in even earlier the next morning.

I used to think it was due to Judy's competitive nature, but years later she explained she didn't want to miss anything and liked to know before I arrived that she had drawn the same conclusion as I had about the previous day's experiments. I thought about Lew Jacobson and my days as a graduate student, when I'd been much the same way. On the rare days she overslept, I'd stay in my car until I saw her bike roll up to the back entrance of the lab.

We recharged and worked off any frustrations of the morning by exercising midday, which could include running, weights, pick-up basketball, or beating me and several other men relentlessly in squash. Then, around 5 p.m., if it wasn't softball season, she'd relax by biking home in the evenings, and then train aspiring cyclists riding centuries (hundred-mile bike rides) on the weekends.

I considered what might be a good research project for her. Judy disliked counting blood colonies under a microscope, so it had to be a virus project. At the time, AIDS was defined as a disease in which certain blood cells, helper T lymphocytes, were infected by HIV and gradually depleted, resulting in the fatal breakdown of the immune system. Several things did not add up. For example, there was a condition called "neuronal AIDS," which didn't make sense because of the absence of T-cells in the brain, as well as the fact that there was only a relatively small number of T-cells which were affected in AIDS patients.

As Judy said to me at the time, "When we know that only one in ten thousand T-cells are infected, the medical community responds by calling it a 'bystander effect.' Clearly, there's another shooter in this scenario, a key immune system target that people are missing." We hypothesized a key was the monocyte macrophage and she decided to study the role of macrophages in HIV infection with the hypothesis that the numerous changes in the brain of AIDS patients resulted from localized viral replication of the monocyte macrophages of the brain called microglia.

Judy started driving from Frederick, Maryland to George Washington University in Washington, DC and back to finish her work in Frederick. If it wasn't rush hour, it was an hour-long drive, but when traffic was heavy it usually took two hours. During that time, we also visited her stepfather Ken in the hospital. I remember him saying, "Frank, shake my hand, you can't do anything else for me." I have remembered and rued that visit a long time.

Ken's nickname for Judy was "Crash" because as he often liked to say, "Judy's mouth always gets Judy's body in trouble." There was another reason for that nickname because once, when Judy got into a serious car accident, she called Ken instead of her own mother because not only did she have glass in her underwear, but she didn't want her mother to worry or kill her for being stupid.

I would always laugh, not realizing I was much the same way, protesting the many injustices at work and yet still being thrown to the curb.

As her written comprehensive exams needed to continue in the PhD program approached, Ken died. I still vividly recall driving to and from the funeral procession with Judy and her asking me to drill her with biochemistry questions. That's the kind of focus she had, even as she was grieving.

Due to her extraordinary efforts, Judy's research thesis was a rousing success. And it resulted in the publication of three important papers.[16] Using a cell line model of monocytoid cells, she found three levels of HIV expression: 1) latent, that is no HIV viral expression; 2) restricted expression of viral RNA, but few viruses produced; and 3) productive expression of HIV.

Latent cells could be activated a year later to produce virus by demethylating agents. Increasing NF-kappa-B activity increased expression from the cells that has restricted its production.

We naturally asked if normal monocytes from AIDS patients had the same ability.

The answer was an overwhelming yes.

Monocytes from AIDS patients harbored latent HIV virus, which was inducible to produce infectious virus during an immune response, leading to T-cell infection and death.

One of the arrogant reviewers for her first manuscript wrote, "Not only do I not believe this paper, but also, I did not believe their last one." This was my first experience, but certainly not my last, that while peer review might be good for technicians or mediocrities that make minimal contributions to the field, it was brutal on innovators.

Well, thirty years later it's clear that monocyte macrophages are a major reservoir for HIV persistence during highly active anti-retroviral therapy and are the major source of virus and viral pathology in the brain.

Allen Goldstein, then the head of the Department of Biochemistry and Molecular Biology, was always very supportive of Judy and her groundbreaking work. (Except when at a department party, she sank a hook shot over him to win a basketball game!) At the private session of Judy's thesis defense, Dr. Goldstein said he wanted to give her a special award, but was told there was no provision which would allow such an award.

* * *

After the discovery of HIV, the dogma surrounding it quickly prevailed over the science of AIDS. All other research projects were dismissed, ostracized, and/or not funded.

AIDS was defined by an antibody positive blood test and any one of several opportunistic infections. However, having one of these opportunistic infections and a negative antibody test meant you were AIDS-free! Anybody who published different data that challenged this dogma was a pariah, who often faced career-altering repercussions.

When there began to be a large number of patients with AIDS-like diseases, but without an HIV antibody positive blood test, the CDC immediately pronounced it was not AIDS, because the patients were HIV-negative.

What a ridiculous circular argument!

You define something that causes a disease but when you find that disease without the cause you defined or that the victims do not meet the demographics, the CDC simply pronounces it is not that disease and makes up a new syndrome for this new idiopathic (unknown cause) disease. Judy likes to say that the idiots don't understand the pathogenesis.

In other words, you had AIDS, but not HIV, and you were not a gay man but a woman. Therefore, you did not have AIDS

Shyh-Ching Lo, who we would meet again decades later, published a series of articles showing the isolation of a microorganism called mycoplasma from AIDS patients.[17] Instead of a scientific discussion on the implications of a mycoplasma infection in people with AIDS, he received essentially nothing but personal attacks.

In 1990, at an AIDS conference, Luc Montagnier, always one to show all the data, stated that he found mycoplasma in about 40 percent of AIDS patients. Dr. Montagnier said, "we thought this one virus was doing all the destruction. Now, we have to understand all the other factors in this."[18] Many scientists castigated Montagnier for squandering his credibility and risking his professional standing, including perhaps a future Nobel Prize. It's incredible to realize a scientist can risk his professional credibility by simply presenting data.

What?

They want a scientist to present only the politically and socially acceptable data. Several scientists have told me that data they do not present is as important as the data they show! Heaven forbid the day that the dogma becomes so important that even its discoverer cannot add or detract from it.

When Dr. Paul Hoffman, working in my lab, made the observation that HTLV-1 could productively infect human macrophages and brain microglial cells, we decided that Judy should do a postdoctoral fellowship in the molecular/virology of HTLV-1, rather than continue in the savage rat race HIV research had become.

At the NCI-Frederick, Maryland, her first choice for a postdoctoral mentor was Dave Derse. For non-scientific reasons, it turned out to be a very poor choice. Better choices might have been non-retroviral molecular biologists. That said, the project of making the first infectious molecular clone of HTLV-1 and the molecular virology she learned has served us both very well, as two decades later, we appreciated the difference between the Silverman infectious molecular clone of XMRV (VP-62) and the natural isolates and defective viruses of the murine leukemia virus family

* * *

As the 1980s were coming to a close, it was a bittersweet time for me. Both of my parents had died without my being able to tell them how much I appreciated them.

In 1981, my father's Hodgkin's lymphoma recurred. I have always wondered that if maybe I'd paid closer attention, then perhaps the outcome might have been different. On his death bed, my old-fashioned father said the family needed a male heir. I'm sure he would have been delighted with our son, born a few years later, and now nearing thirty-five years old. He is and always was a joy.

In 1987, while Sandy and I were at a meeting in London, my mother passed away. I lament that we would never have the opportunity to understand each other.

Sandy and I had been talking about moving to a more family friendly environment like Frederick, Maryland, rather than the rich and posh neighborhood of Bethesda where we were living at the time.

In 1989, Sandy had the chance to join the laboratory of molecular oncology, headed up by Dr. Takis Pappas, at NCI-Frederick. We looked around for a property and found one we really liked in a planned community. We bought the largest unzoned lot with direct views of the lake. I had to petition the Frederick zoning commission to get the property zoned for residential use. Sandy and I had plans drawn up for a contemporary home and found an architect. Friends like Judy made useful suggestions.

Our place in Bethesda sold quickly, even though we listed it for more than our realtor suggested. We stayed in Judy's apartment in Frederick until our home was ready. Judy had been spending most of her time with her

mom after Ken's death, and her commute to complete her research thesis
was much shorter from northern Virginia.

We loved the bright and airy place we designed, and it was a short
fifteen- to twenty-minute drive to work. It was a great place to raise children,
providing easy access to events in either Washington, DC or Baltimore.
We lived there from 1990 to 2014, until we moved to Carlsbad, California,
located on the Pacific Ocean.

CHAPTER EIGHT

Be Careful What You Wish For

The section chief at NCI is the best job in science;
Laboratory chief has all the "crapola."
　　　　　　　　　　　　　　　　　　　　—Mike Potter

Science does not have a moral dimension. It is like a knife.
Give it to a surgeon or murderer. Each will use it differently.
　　　　　　　　　　　　　　　　　　　　—Wernher Von Braun[1]

By the time we moved into our new lakeside home in Lake Linganore, we felt we could have a normal family and professional life.

In 1992, as a result of Judy's thesis and my discovery of TGF-beta as a master regulatory switch of blood stem cells[2], my laboratory focused on the innate immune system, retroviral surface proteins, viral pathogenesis, and normal cytokine regulation of inflammation, immunity, and hematological stem cell growth. Throughout the 1990s, my lab varied from eight to twelve people.

As chief of the Lymphokine (proteins released by lymphocytes) section of Joost Oppenheim's Laboratory of Molecular Immunoregulation, I was only responsible for the research of those eight to twelve people. I was responsible for integrating the data/knowledge generated by those scientists and technicians in my section. It was a challenging and exciting position.

Joost and I would occasionally argue about many topics. Once, he stopped me in the middle of an argument to ask if I was Jewish.

I said, "No, why?"

He replied, "You argue like one!"

I took it as a compliment.

In 1992, I was awarded the Distinguished Service Award, the highest honor the US Department of Health and Human Services can award. The award read as follows: *In recognition of the fundamental co-discoveries; Interleukin-2, the first human leukemia virus, and for discovery of hematopoietic regulatory activities of transforming growth factor beta.*

In addition, I had the honor of organizing a human retroviral symposium in Genoa, Italy. That honor did not come without a price, as scientists can be just as demanding as any bad tourist. Some wanted free airplane tickets for their spouses, first class tickets, or to change their speaking order or time. For me, it was only about communicating the knowledge.

Of course, it did not hurt to be in a beautiful castle in Genoa, Italy and eating fine food. But some scientists seem to believe it's not about the knowledge, the beautiful surroundings, or good food, but about what science can do for them. The meeting went off smoothly and I was able to get the job done. It turned out to be a great experience, and I learned valuable lessons about the politics behind the organization of scientific meetings.

In 1992, I was made chief of a new laboratory, the Laboratory of Leukocyte Biology. I'd like to believe the laboratory was created because our team had made so many paradigm shifting discoveries. However, it was quickly apparent that there were other agendas.

It probably would have been better if I'd taken over an existing lab with some established administrative and obvious research goals. I should have heeded fellow scientist Michael Potter's warning that "the section chief at the National Cancer Institute is the best job in science. Laboratory chief has all the crapola."

I had no choice in the decision of which independent investigators would be in my lab, regardless of whether their research was relevant to the mission of the Laboratory of Leukocyte Biology. My current boss wanted me to hire two young investigators; one scientist was a natural choice, while the other, a former technician, was a poor fit. George Vande Woude, who would become my boss, wanted me to take another scientist because there "weren't many Dutch Jews in science." This was a terrible choice, as this scientist would go over my head and report directly any complaints to George. Finally, Dave Derse begged to let him join my lab because he despised his current boss. I also felt I owed him because he'd agreed to be Judy's post-doctoral mentor, so I agreed. These additions greatly added to my administrative burden and associated stress.

Ironically, at the same time in 1993, Sandy's lab chief, Takis Papas, departed for a cancer center in South Carolina and she was appointed acting lab chief. Dr. Pappas left without performing any of the administrative tasks necessary for closing a laboratory. Of course, they chose a woman to clean up the mess a man left behind. But at least her dissatisfaction was shorter-lived than mine.

There were some nice opportunities that came with supervising a laboratory. I had a chance to invite seminar speakers whose work I really admired. I was able to invite established investigators like Steve Bartelmez from Seattle and Shin Kang from South Korea to spend time in the lab. Steve and I have now collaborated for almost four decades, making seminal discoveries regarding TGF-beta regulation of stem cells, just as Judy and I have worked for nearly the same amount of time, making seminal discoveries in human retrovirology. It seems that never a day has gone by without Steve or Judy challenging me to consider their perspective.

And neither of them ever challenge me quietly.

* * *

Sports was always a fun way to relax from the pressure of the lab.

The Baltimore Orioles opened Camden Yards in 1992, and Judy and I bought two seats in the twenty-nine-game season ticket plan. The plan was set up so we got to see one game of every visiting team that came to town. Either Judy and I would take in a game, or I'd take my son, or Judy would take one of her two nephews. The best evening of all was the September 6, 1995 game where Cal Ripken broke Lou Gehrig's record for the most consecutive games played. The next week we saw Eddie Murray hit his five-hundredth home run.

As I mentioned before, Fort Detrick had an intramural slow pitch softball league and Judy would recruit and manage the team. We were always in the playoffs and often won the title.

Judy had a little sister in the Big Brothers, Big Sisters program, Karen, who eventually married a local disc jockey, Dan Stevens. Like us, Dan started his disc jockey job before 5 a.m. and would occasionally dedicate his opening song to us, saying we were already at work trying to cure cancer. It was such a pick me up and it made us want to keep working hard.

Around this time, Sandy and I began a habit of taking our son to international meetings to which we were invited.

We eventually took him to meetings in Brazil, Jamaica, Canada, Germany, Sweden, Morocco, Japan, and South Africa. In South Africa, we saw one of my personal heroes, Nelson Mandela, and took a safari in Kruger National Park.

If possible, I think everyone should travel outside the country, as it is a great education.

Sandy and I received many invitations to speak at international meetings. Among my favorites was a presentation at a symposium in the reopening of the Curie institute in Paris in 1995. The speaker podium in the conference room had been made from Marie Curie's laboratory bench. Being told that it was actually her bench, I slowly backed away from the podium (radioactive, maybe) while speaking. Another was being made a fellow of the University of Bergen during Sten-Eirik Jacobsen's doctorate celebration.

* * *

Things were changing at the NIH after the 1987 announcement by Harold Varmus that the French scientists would get more of the HIV test kit royalties.

This put subtle pressure on Gallo to leave the NIH.

I shall quote from the book, *Science Fictions: A Scientific Mystery, A Massive Coverup, and the Dark Legacy of Robert Gallo*, written by Pulitzer Prize-winning *Chicago Tribune* writer, John Crewdson. After a three-year investigation of Gallo's actions by the Office of Research Integrity at the NCI, he resigned in 1994, and after a search, started the institute of Human Virology at the University of Maryland. In the closing paragraphs of Crewdson's book, he tried to capture the character of my former boss. Crewdson wrote:

> Robert Gallo played the game of Big Science better than it had ever been played before. He had published, by his own count, over a thousand scientific papers, including the ones he had actually written. His *curriculum vitae* listed eight single-spaced pages of prizes, honors, and awards. He had become the most famous AIDS researcher in the world, treated like scientific royalty on three continents.
>
> But for the tens of millions of dollars that flowed into Gallo's lab the taxpayers had gotten precious little beyond HTLV-1 and T-Cell Growth Factor . . .
>
> Being wrong in science is hardly a sin. Scientists are wrong every day, and in a curious way mistakes are what pushes science forward. What set

Robert Gallo apart was his profound disinclination to acknowledge his mistakes, preferring to ignore them, insist they hadn't occurred, blame someone else, or propagate outlandish fictions that only confused science further and slowed its forward march.[3]

I thought Crewdson did an accurate job of capturing the duplicity of my former boss. But is it enough to point out the flaws of the man, as well as his good friend, Anthony Fauci, who has messed up the coronavirus crisis every bit as badly as Gallo and Fauci did the HIV-AIDS epidemic?

People like Gallo and Fauci thrive because of a system put in place by President Richard Nixon, in which a few leading government scientists were given unprecedented control over funding decisions, while at the same time creating a subservient class of researchers who must vie for their approval.

During the same time period that Gallo was being driven out of government service, three of his associates were getting into trouble over the misappropriation of government funds.

Prem Sarin, Gallo's second in command who I claimed took a cell line I developed and published it as his own, was convicted of monetary abuses:

A prominent AIDS researcher was sentenced today to two months in a work release facility for embezzling $25,000 earmarked for AIDS research at the National Institutes of Health.

Prem Sarin, 57, of Gaithersberg, who faced up to 20 years in federal prison, pleaded for leniency, citing a need to continue his research.

Sarin contended in court papers that he is a victim of a vendetta by a House subcommittee investigating scientific claims of Sarin's former boss at NIH, AIDS researcher, Robert C. Gallo.[4]

Get the argument? I'm arguing that my research is too important for you to put me in jail, and I'm also saying this is all because of an evil "vendetta" by a House subcommittee.

Zaki Salahuddin, who Gallo thought put the viruses in HL-23, resigned over financial abuses:

A biologist working for the National Cancer Institute in the U.S. has been suspended following claims that he has been steering government funds toward a firm partly owned by him and his wife. The firm has provided equipment and materials to the laboratory in which he works.

The case has attracted widespread attention because the biologist, Syed
Zaki Salahuddin, works in the NCI's laboratory of tumor cell biology, which
is headed by Robert Gallo, co-discoverer of the AIDS virus.[5]

Importantly from my perspective, the research by my laboratory on the
developmental regulation of blood cells was proceeding excellently. Our
team, consisting of Keller, Jacobsen, Dubois, and others, was pioneering the
role of cross-talk between cytokines in regulating the surface expression of
cytokine receptors. This was important in producing the additive effects of
positive-acting cytokines on cell function.

Also, this mechanism was important in controlling the interactions
between opposing factors that inhibited or stimulated cell functions,
depending on the context. These results and Sandy's work on the interactions between viral surface proteins and cell surface receptors in viral
pathogenesis greatly informed our viral work. Sandy's work showing that
changing just two amino acids in a viral envelope protein could change the
pathogenesis from a leukemia to a neurological disease was a real breakthrough in our understanding of how subtle sequence changes in retroviruses can affect disease phenotype.

The lab's research into human retroviral pathogenesis was going equally
well. Judy was working on successful projects as a postdoctoral fellow, such
as developing the first molecular clone of HTLV-1, which was useful in
determining the gene expression required for malignant T-cell growth, and
the role of the naturally occurring defective (incomplete) viruses during
viral infection. As a technician, her projects included the role of methylation
(which alters gene expression) in retroviral life cycles; anti-viral mechanisms
of type 1 interferon; and other cytokines. Judy and Dave Derse gave presentations at the 1992 Cold Spring Harbor RNA Tumor Virus meeting on the
characterization of these defective viruses.

Scientific progress seemed to be running smoothly, but in the late 1990s,
a turn for the worse was brewing. Derse was becoming irritated that Judy
was wearing two hats. Remember all of her salary was being paid by my lab
budget and she still had to conduct scientific studies outside of her postdoctoral studies to receive her salary. Thus, she was not able to be fully dedicated to his instructions. But he was able to recruit two other fellows who
were. I always felt that Derse did not want to share the HTLV-1 podium
with me and that was the root of the problem.

Whatever the reasons, Derse made enough loud and unfair complaints
about being in my lab that he received permission to find a new lab. While

these events were occurring, Derse had submitted a paper on defective viruses with Judy as senior author. After Derse left my lab, he took Judy's name off the paper, a thoroughly unethical act. In science, credit and publication are the coins of the realm, and nobody needs it more than a junior researcher.

I was completely disgusted by this theft of Judy's work by Derse.

* * *

I'd published many papers concerning the regulation of blood cell development, concluding with Steve Bartelmez's paper on regulation of stem cells with TGF-beta.[6] The use of the anti-TGF-beta antibody was a crucial reagent in the success of any investigation.

I got a call from a former colleague, now working at Biogen, who told me that Biogen had bought the antibody, and I was now forbidden to use it in my research. I then produced a signed material transfer agreement which allowed me to use the antibody. I told him further that the antibody producing cells were available from American Type Culture Collection, for all investigators to use.

The cells were removed from the American Type Culture Collection within forty-eight hours after my phone call. I was then informed I would get no more TGF-beta reagents, particularly the humanized antibodies they were making.

So, at a key point, after almost a decade of studies, we needed to find a new technology, particularly to translate our findings from mouse studies to humans. Bartelmez began a wonderful collaboration with Pat Iversen, who was pioneering a new technology using phosporodiamidate mortpholino oligomer (PMOs). PMOs are safe, synthetic antisense molecules which bind to RNA molecules in a sequence-specific manner to prevent a given protein from being made. He made several PMOs against TGF-beta that were very efficient.

Pat Iverson, using PMO antisense molecules as antivirals, has shown that PMOs were very effective against the lethal Ebola virus and SARS-1 coronavirus, as well as several other viruses.

Why nobody was listening and didn't try it against COVID-19 is beyond me and a clear example of the myopic nature of science.

The success of the Derse revolt emboldened other members of the lab to voice that they wanted to leave the lab. The report from the 1998 site visit contained stunning requests to leave.

It was stunning because none of these people ever told me they were unhappy. The request from a female investigator was a shock because I had spent lots of time giving her the ability to get tenure, when the site visit itself had recommended eliminating her position. Ironically, the only investigators who did not complain were the ones who had performed excellently in the site visit. To this day, I do not know what I did so wrong to warrant the horrible reviews from those I'd attempted to help.

Furthermore, three other scientists moved on from the lab to have very good scientific careers.

Well, at least hard-working clever Judy didn't leave the lab, at least not voluntarily. She was forced out by several people, the most vociferous of which were the Derse lab members who stole her work.

Judy published a seminal paper that HIV infection upregulated DNA methyltransferase, which caused de novo methylation of the gamma interferon gene, which downregulates the production of gamma interferon.[7] This mechanism is one way the virus evades T-cell immunity. Furthermore, she showed that infection with an HIV virus which could not integrate, and therefore could not be infectious, was still able to stimulate de novo methylation, suggesting that a virus can cause pathogenic changes without productive infection.[8]

In January 1999, Judy was made head of the Lab of Antiviral Drugs Mechanisms. I felt it was done because it was thought that separating us would make us both weaker. I believe the NCI hierarchy fully expected and wanted her to fail. They also believed it would make my lab weaker and less creative.

One of Judy's hires at the Lab of Antiviral Drug Mechanisms was from Shanghai and did not want to take orders from a woman. Judy would write up the experiments she needed to be done, and then I'd give them to the doctor, who would do the work.

Needing to find a replacement for Judy, I hired an experienced retrovirologist who was working in Baltimore. Gallo was coming to work at her university, and she wanted to get out of there as quickly as possible.

With the cloning of the chemokine receptors, it became obvious that chemokine receptors were a cofactor for HIV cell entry. Peptide T specifically inhibited viral strains which used CCR5 receptors for entry into the macrophage, microglia, and CD4 T-cells, but had no effect on lab-adapted strains using CXCR4. This was vindication we were right.

But as usual, nobody was listening. We hoped there would be other applications for this reagent.

* * *

Being absolutely sure that Peptide T would be a useful reagent, I wrote a grant between Candace Pert, Michael Ruff, and their company to study how to make Peptide T a commercially viable product.

With reviews straight out of 1986, my request to study Peptide T was denied.

In 2001, Judy permanently left Frederick, the NCI, and the Lab of Antiviral Drug Mechanisms. She accepted a scientific research position with EpiGenX Pharmaceuticals in California. This move allowed her to pursue her three loves; methylation, viruses, and David Nolde, whom she married on October 7, 2000. Her instincts were right about methylation.

It is my feeling that epigenetic therapy to eliminate the pre-metastatic niche before tumor cells can use these niches to metastasize to other body parts will be the next major breakthrough in cancer treatment. One of the mechanisms involved is fascinating to us. The epigenetic therapy induced endogenous retroviruses expression which stimulated the production of interferon-alpha in the host, which is part of the immune response to eliminate the pre metastatic niches. That is to prevent the cancer from spreading.

For her farewell party, several of her close girlfriends and me (the designated driver) celebrated at Tauraso's restaurant (our usual spot for lab Christmas parties). Nick Tauraso, a former virologist turned sommelier/restaurateur, paired excellent wines with food. I drove all the women home, with Judy being the last stop.

As I was walking to my car, I heard a loud crash, which made me turn around. Judy had fallen over the railing and landed on top of the Japanese maple, recently planted by her brother-in-law. I rescued her and made sure she got to her apartment without further incident.

All of her belongings had already been shipped to California, so that next day she flew out wearing the same pair of jeans she'd worn the previous night, which now had a large rip in them. We never told Judy's brother-in-law how his Japanese maple had been destroyed.

A result of my laboratory's next site visit in 2002 was the recommendation that my laboratory be closed. Maybe it was inevitable. Three of the four independent investigators had demanded to leave the lab and the fourth had already left the lab.

It was easy for the site inspectors to claim the lab was in decline. But from my perspective it was a different story, and one all too common in government. Having two hands tied behind my back, I have never felt I was given a fair chance to succeed. It was a bitter pill to swallow, considering the years I'd dedicated to government service.

CHAPTER NINE

Millennium Changes

I worry that, especially as the millennium edges nearer, pseudoscience and superstition will seem year by year, more tempting. The siren song of unreason more sonorous and attractive.

—Carl Sagan[1]

Two decades into the millennium, several studies have demonstrated that when compared to other developed countries, more Americans prefer dictatorship over democracy.

There could be many reasons for this, but one is the complete failure of the public health apparatus, the FDA, the CDC, the Environmental Protection Agency (EPA), NIAID, NIH, and the HHS, who are charged with making the regulatory decisions to protect our public health. Their mixed messages in any crisis, the many decisions that enrich their friends and regulators, and the utter contempt with which they treat the public lead to a crisis of confidence. Most citizens, including those who research these issues deeply, do not know who to believe.

As I reached another crossroads with my laboratory torn asunder, could I continue and rebuild my lab again? With all the corruption, I knew that no matter what, I must live my life within the bounds of my value system. My obligation was to continue to simply focus on the science as I have always done.

Luckily, as long as I had a budget, I could work on projects without having to justify the funding, at least until the next four-year program review.

Elsewhere, as a Nobel laureate wrote on the 2014 death of Fred Sanger, a two-time Nobel Prize winner:

> A Fred Sanger would not survive in today's world of science. With continuous reporting and appraisals, some committee would note he published little of note between insulin in 1952 and his first paper on RNA sequencing in 1967, with another long gap until DNA sequencing in 1977. He would be labeled as unproductive, and his modest personal support would be denied. We no longer have a culture that allows individuals to embark on long term and what would be considered risky projects.[2]

Most of the scientists who perform experiments agree with me that this change in culture was led by the rise in dominance of the bureaucrats/physician-bureaucrats who are themselves fundamentally unproductive, and that this is a serious long-term threat to science. However, the correct use of science can still be a "candle in the dark."

The exploration into uncovering the causes which are sure to be multi-factorial in the initiation and progression of chronic diseases has always intrigued me. I would continue to investigate the role of the innate immune system in the pathogenesis of retroviruses, particularly the cells which regulate type I interferon, the body's natural antiviral, and the use of transient inhibition of TGF-beta in stem cell therapies in chronic diseases.

The concept that macrophages could support productive, restricted, and latent expression from Judy's thesis persuaded us that we needed to examine the innate immune system to fully understand pathogenesis and develop therapeutic strategies. The innate immune system (consists of rapid physical, chemical, and cellular nonspecific defenses against pathogens) is the first line of defense against invading pathogens.

In the innate immune system, sensors like pattern recognition receptors (PRRs) detect specific viral components such as viral RNA, DNA, or intermediate products which stimulate production of interferon alpha/beta and activate other immune cells. Recent studies show interferon lambda directly stimulates restriction factors that inhibit HIV replication and also upregulates intracellular expression of type I interferon. It also shows that interferon has both extracellular and intracellular antiviral mechanisms which differ. To use interferon alpha therapeutically, one must control the potential detrimental effects.

Never having been comfortable with the fact that cell-free HTLV-1 virions are poorly infectious *in vitro*, I asked my new staff scientist to examine infectivity in innate immune cells. She showed HTLV-1 could efficiently infect myeloid and plasmacytoid dendritic cells (DC) and even more importantly, could rapidly and efficiently transfer infectious virus to autologous primary T-cells which became infected.[3]

We also found that both types of dendritic cells could act as a true viral reservoir. The first cells to be infected *in vivo*.[4]

These studies suggested to us that impairment of the dendritic cells function after HTLV-1 infection plays a role in pathogenesis.[5] Emphasis was placed upon HTLV-1 induced alteration of type I interferon shown by several investigators. It was our hypothesis that the development of acute T-cell leukemia (ATL) is a struggle within the body between HTLV-1 and type I interferon production. The HTLV-1 p30 protein is required for infectivity *in vivo* and for dendritic cells *in vitro*, but not for T-cells.

The p30 protein was shown to dampen interferon responses in monocytes and dendric cells, representing an early step for HTLV-1 infection. Interferon type I and AZT treatment of acute T-cell leukemia can result in complete remission.[6] Again, this suggests that exogenous type I interferon, in contrast to endogenous interferon type I, can act therapeutically in HTLV-1 infection.

What about other viruses?

Kaposi's sarcoma associated herpes virus, in which B lymphocytes are the major infected reservoir in living organisms, cannot be infected by the virus in lab culture. We demonstrated that the dendric cells mediate infection of primary B-cells.[7] We point out in the interferon chapter in Part Two that, for some unknown reason, Fauci, despite praising interferon use in AIDS patients and NIAID running a successful trial in 1989,[8] failed to recommend interferon and AZT trials when AZT was the only approved drug and obviously was working alone.

Why is Fauci again ignoring using interferon for COVID-19, where an imbalanced host response, defined by low levels of type I and type III interferons (antiviral defenses), next to elevated chemokines and IL-6 (inflammatory marker) are driving the viral inflammatory response? It is clear as several papers have shown that there are benefits to exogenous interferon treatment using either interferon alpha or lambda.[9]

The second project involved a collaborative effort between Steve Bartelmez, Pat Iverson, and me, involving the *ex vivo* inhibition of endogenous TGF-beta in purified hematology stem cells (HSC). The astonishing results permitted complete functional transplant with as few as sixty HSCs, instead of a thousand cells, which could rescue mice from lethal irradiation.[10]

Both the anti-TGF-beta antibody and PMO antisense TGF-beta treated HSCs produced accelerated engraftment, resulting in high levels of donor cells, and a durable graft. Also, never seen before, the early donor cells were neutrophils, which protect a transplant against infection. This was truly miraculous and paradigm-changing.

After twenty years, we can still not get it published in a good journal because the reviewers don't believe it. It's not religion. It's science. It was important enough that I wanted to compromise with the reviewers to get it published, but Steve never wanted to do this. Without telling me, Steve submitted it to a journal (*Stem Cell Research and Therapeutics*), which few could ever read because it is not available online.[11] If a scholar's obligation is to share knowledge, in this instance we have failed dismally. It is published, but nobody is reading it.

To continue these cell therapy studies, we needed to show that this therapy would work on human cells. Steve wisely chose diabetic retinopathy, a common cause of vision loss and blindness. Healthy human HSCs can go to the retina and help repair vascular leaks, but HSCs of diabetics cannot do this vascular repair. We demonstrated that human specific PMO antisense TGF-beta can reverse the dysfunction of diabetic HSC so that they are now able to repair the endothelium in the retina.[12]

Since the only clinical therapy is anti-VEGF, which fails 50 percent of the time, I believe this additional therapeutic approach should be an attractive approach for those patients who need a therapy. I have a great deal of frustration that none of these observations have led to clinical trials despite a decade of effort.

I believe *ex vivo* manipulation of stem cells have a bright future, if the funds are provided for the appropriate investigations.

* * *

Many of my collaborators and friends seemed to be passing out of my life, and I didn't understand why.

One of the most lamentable losses was Dr. Kendall Smith. I always praised his world-class work on the biochemical mechanisms of IL-2, which he originally called "Frankie's factor." He often invited me to his lab and I'd stay in his home, having long conversations about science and enjoying his wife's amazing French cooking.

One of the best AIDS meetings we ever attended was in Reims, France in April, 1995. It was the first AIDS meeting that acknowledged the importance of cytokines in HIV pathology and treatment. Kendall knew French cuisine, so Judy, Kendall, and I went to an expensive restaurant during the conference, at which the Prix Fix meal was an all-truffle menu. None of us knew that the menus left the last digit off the prices, and that the women's menu had no pricing at all.

Knowing Judy was still a graduate student on a technician's salary, Kendall and I tried to eat less expensive meals and give Judy the opportunity to try the truffle menu. Even with the truffle ice cream, the price appeared to be $150.

Judy gave Kendall her American Express card to pay for the bill and Kendall took care of everything since he spoke French fluently. When Kendall returned to the table, he seemed uncomfortable.

The next day, Judy and I were on the bus going back to Paris with the other conferees, who had also gone to that restaurant. They asked if anybody had gotten the truffle menu and were surprised when Judy said she had. They asked if it was good.

Judy said it was, but the ice cream tasted like dirt.

Then Judy wondered what the big deal was and pulled out the receipt. Her eyes nearly popped out of her head when she saw the bill for the meal was $1,500.

We thought there must be some mistake so as soon as we got to Paris we walked all the way to the American Express office and asked them to call the restaurant to inquire if the bill was correct.

We overheard the clerk assure the restaurant we loved the meal and just wanted to make sure we'd given the wait staff the appropriate tip. We almost choked.

To this day when someone asks if a restaurant is expensive, we reply, "Just a truffle."

At his sixtieth birthday party in 2002, Kendall graciously proclaimed I taught him how to do science. But the good feelings were not to last.

In 2007, Jo Oppenheim was asked to write an article in the *Journal of Immunology*'s "Pillars of Immunology" section, celebrating the discovery of IL-2.[13] He did not ask me to coauthor or review it prior to publication. It minimized Kendall's important role in IL-2. A few days after its publication, I received an email from Kendall, saying, "Et tu, Brute?" He thought I'd stolen credit from him.

Kendall and I have never spoken since that email. Is the amount of credit received for scientific discoveries more important than friendships? Gallo called me many times over the years complaining about

Kendall demeaning our contributions and claiming too much credit for himself.

* * *

Several times, Gallo's subordinates invited me to give a lecture in Gallo's new virology institute at the University of Maryland to give a lecture. When I was finally asked about my constant refusals, I responded, "I only want to piss on his grave. But ever since I was in the service I've never wanted to wait in long lines, again."

I did not receive any further invitations.

Much later I heard that the Nobel Prize committee would have considered Gallo if Bernie Poiesz and I had made peace with him. We never knew if this was true, but if so, it is poetic justice.

When Gallo did not get the Nobel Prize, most of my friends were surprised that I had mixed feelings about the decision. It would have been a de facto recognition of the work Bernie and I had done in his lab.

But Gallo's lack of character spoiled it all for us.

* * *

My greatest pleasure in the early part of the twenty-first century was teaching, practicing, and watching our son, a left-handed pitcher/outfielder, play traveling league baseball.

One of my top thrills was watching him pitch at Saint Mary's Home for Wayward Boys, where Babe Ruth had pitched as a young man. World Series tournaments in Tennessee and Mississippi in the summer were miserably hot, but we remember them fondly. Our son graduated from high school in 2004, and from the University of Virginia in 2008, which had one of the few surviving student-run honor systems in the country. I'd hoped he would play baseball in college, and possibly in the pros, but he preferred the family business of science.

In 2007, I was elected to the Senior Biomedical Research Service, for which I had been nominated and denied since 1994. (Sandy had been elected in 1996.) Although the position was an honor with a higher salary, it had a downside. One had to give up their civil service retirement protection to accept the honor, which would later become an issue for us both.

In 2009, I was the first coauthor and Sandy was also a coauthor of the paper in *Science*, showing isolation of XMRV, a mouse retrovirus, from the blood of patients with chronic fatigue syndrome (ME/CFS).[14] We did not use the word "cause" in the paper, as the isolation of a virus from a disease population does not usually involve showing causation. (The first paper on HTLV-1 and HIV did NOT prove causation.) Yet everything else about the XMRV discovery was ignored, and Judy was accused of saying it caused chronic fatigue syndrome, which she did not. Since Judy and Kent's first book details these events, I will provide my own recollection of events which followed the publication of that paper.

First, the Department of Health and Human Services called Judy and I the day after the paper was published in 2009 to say its first priority was to determine if the blood supply was contaminated. I suspected this was a knee jerk reaction by the government to their failure to act quickly on this question during the HIV/AIDS epidemic. But, of course, we both enthusiastically agreed to participate, happy the government was apparently going to do the right thing.

Second, John Coffin gathered a group of intramural investigators to develop diagnostic tests, many reagents, and possible therapies, to the tune of about a hundred thousand dollars, so that the NIH and Coffin could be the leaders in studies of this new human retrovirus' role in human disease.

Third, the following year, 2010, Harvey Alter (now a Nobel Prize laureate) and his long-time collaborator, Shyh-Ching Lo, published a confirmatory study in the *Proceedings of the National Academy of Sciences*, whose findings showed an even higher association of XMRV-related sequences with chronic fatigue syndrome patients.[15] That paper, as Alter said, not only confirmed our study but detected variants; that is, there were other gamma-retroviral sequences in the blood of patients with chronic fatigue syndrome and heathy blood donors.

However, Harold Varmus, at the time the head of the NCI, quickly said that it was not a confirmation because it was not the same strain or sequences. The confirmatory studies for HIV and HTLV-1 did not require the exact same strain of the virus! Varmus quickly said in the press that the Lo/Alter study only confirmed the presence of murine viral sequences in human blood samples, not the presence of infectious and transmissible viruses, and thus our work had not been confirmed.

Next, several "peer-reviewed" studies, one in as little as three days after our paper was published, were rapidly published saying there was no evidence of any murine viral sequences. John Coffin quickly declared all the positive sequences were lab contaminants. In fact, rumors were leaking from his Frederick lab that this was the case before having the courtesy of telling me, as I was a collaborator in his XMRV group and on these publications. As usual, gossiping about credit at NIH brings out the worst in collaborators. Next, several more studies after the Lo/Alter paper was published concluded there was no evidence of mouse viral sequences.

Judy and Harvey Alter, as well as their collaborators on these publications, have steadfastly maintained to this day the absence of contamination in any of their laboratories. Let me give you my perspective so there can be no mistake: there was NEVER any evidence of contamination in the Mikovits or Alter laboratories.

However, the increase in the number of papers publishing negative results caused the National Heart, Lung, and Blood Institute (NHLBI) to change the study design of the blood working group to assess the risk of blood safety of not only XMRV but all MLV variant sequences.

Dr. Michael Busch, director of the Blood Systems Research Institute, was supervisor of the study and senior author on the September 22, 2011 *Science* paper about the blood supply. In phase 1a, a small pilot study panel to test the specificity and sensitivity of the PCR tests of each lab, the Coffin Lab's PCR test reported all samples as negative. Judy's PCR test confirmed the results in the *Science* paper, detecting all four patient samples as positive and the four controls as negative. The CDC's PCR test also got positive results in all four patients, but not in controls.[16] Phase 1a used fresh plasma, processed from blood, two and four days after being drawn, from ME/CFS patients. These were the same patients in whom XMRV viral protein sequences, and in two cases viral isolates, had been shown in our 2009 *Science* paper.

As you might imagine, Coffin said that Judy's lab and the CDC PCR tests had sensitivity issues. Coffin felt that these PCR results were false positives. Coffin did not consider that *his* lab's PCR test had sensitivity issues. This is, it could not detect low levels of XMRV expression.

That is what Judy meant when she wrote in an email (included in this book), that the negative cutoff on Coffin's test was so high it would not find a willing Roman in a whorehouse. Coffin "fixed' the PCR primer issues and the positivity reported by the CDC disappeared. The CDC never explained. The Abbott serology test, which had previously reported a few positives, was now reporting all samples as negative.

Judy objected to the *Science* paper senior-authored by Dr. Michael Busch, its conclusions and title, because the study was not statistically powered to be valid as a confirmatory study of the virus in CFS samples. Since the goal was to ascertain the safety of the blood supply, both Judy and I agreed we would not coauthor the paper unless the title and the conclusions were changed.

Judy called me very concerned because Simone Glynn and Michael Busch had called her on the phone. She said Harold Varmus was also on the line and said that if Judy and I did not coauthor the paper, Sandy and I would lose our retirement, as Varmus would claim the 2009 *Science* paper was fraud.

A decade later, I am still sorry I did not call Simone Glynn's bluff.

But since we did not have the civil service protections because of our membership in the Senior Biomedical Research Service, losing our retirement for more than seventy-five years of service, was a scary but unlikely possibility.

I was additionally sorry because Judy had given a talk at the New York Academy of Sciences on March 29, 2011 saying that the INTERCEPT system developed by Cerus worked on any XMRVs in the blood supply. Judy had spiked 148 platelet samples with the infectious molecular clone of XMRV (VP-62), and after treatment with the INTERCEPT system, all showed no detectable infectious virus. So, Judy had already shown that blood supply could be safely decontaminated and protected. Naturally, Cerus is still using this system.

So, they could have told the truth. They had the technology to decontaminate the blood supply and prevent future contamination of the blood supply. Why was it necessary to instead destroy Judy's career and reputation?

* * *

In June 2011, at a retrovirus meeting in Leuven, Belgium, Bob Silverman, whose lab did most of the PCR for the *Science* paper, told me that he and his collaborator, Joy Das Gupta, were sorry but found all the positive samples were positive for the VP-62 plasmid he had created, and none of the healthy negative controls in Figure 1 were positive. That's highly improbable, if not statistically impossible if the samples were handled the same.

Silverman said he was going to write a retraction in *Science*. When we received the letter that he was going to send to *Science*, the implications were that it was not his lab's fault.

My head nearly exploded because it meant to me that the way they did the experiments, all PCR sequences were excluded except VP-62—the molecular clone of XMRV—thereby missing all other possible strains or variants, which probably suggests that Mikovits and Alter labs were right in finding variant sequences and that the Silverman lab may have treated patients and normal samples differently, which never should be done.

This is a lesson for all scientists.

Be careful with whom you collaborate. Make sure that they are men and women of integrity who honor a scholar's obligation to show all the data and communicate it regardless of the cost.

Judy's treatment after the publication of the 2009 paper finding XMRV was unbelievably cruel.

First, the endless greed of the Whittemores to capitalize on the discovery made relations difficult for her within the scientific community. I was required to attend NIH meetings with the Whittemores where they tried to get money without peer review from the NCI and NIAID, and, with US Senate Majority Leader Harry Reid, to get money through Congress to bypass the NIH. It's my opinion they started to sell an unproven PCR test for XMRV in patients though a spin-off company, VIPDx, using NIH grant money to fund it, as Judy and I discussed at the time.

Judy pleaded with them not to sell the test until the science was resolved.

I personally witnessed Harvey Whittemore grab Judy by the neck and threaten her during discussions. Judy and Kent detailed these incidents in their first book, *Plague*.

In September 2011, Judy was fired from the Whittemore-Peterson Institute (WPI). In my opinion, this was so the Whittemores could keep all their ill-gotten revenues from their unvalidated PCR test for XMRV, company collaborations, and NIH grants with Judy as principal investigator. Therefore, it was necessary to scapegoat Judy.

It was obvious to me that Judy had been "home-towned" from the beginning of the civil and criminal cases by the Reno criminal justice system, because of the controlling influence of the Whittemores.

The savage bail hold placed on Judy as a flight risk when she was living in her own home was an abuse of power and resulted in her being confined to jail for five days.

The bail bondsman who eventually provided her bail said a convicted drug dealer would have gotten out quicker than Judy.

Not only was she prevented from preparing a defense for herself, but she was ordered to surrender herself in Reno for rebooking.

Nothing about the treatment of Judy convinces me that justice exists in America for figures at odds with the power structure.

I first talked to Ian Lipkin after Anthony Fauci announced that the government was going to sponsor an XMRV replication trial, directed by Lipkin, applauded in the media as the "world's best virus hunter." I must admit that I'd never heard of Lipkin and there was little in his curriculum vitae which justified the media praise.

All Lipkin has ever done, in my opinion, is to misconstrue the truth, make promises he never intends to keep, and then complain about not having enough research funds.

While Judy was in jail, Harvey Whittemore told Ian and me that Judy would not get out of jail until she signed an apology that admitted that she was guilty of fraud in the XMRV CFS studies and returned the notebooks. Then she was out of jail and the notebooks returned. Those notebooks, though she would have legally had copies, were not in her possession the day she was jailed, November 18, 2011.

But now the NIH wanted to finish the demolition that the Whittemores started. On November 15, 2011 (three days before Judy was jailed), it was decided that Judy could participate in the Lipkin study in my laboratory in Frederick.

On November 30, it was decided that she could not come to my lab in Frederick to do the experiments but could work as a consultant from a distance. Virus isolation by phone? Judy was good, but not that good.

On December 2, 2011, the two-pronged attack delivered two crushing crescendos. In Reno, Judy had to be present in the viewing of the returned lab notebooks to detail where she went, as well as the time, place, and chronological order of each notebook. Still, the eternally corrupt Whittemores thought there had to be more to steal. They needed to have all of Judy's information to claim as their own to be forgiven their crimes and keep the NIAID grants, which they receive to this day. (As do several other University of Nevada Reno professors who participated in the destruction of Judy's career.)

At the same time, Judy and I received an email from Lipkin's program assistant concerning arrangements to attend a December meeting to discuss the Lipkin study. The email read in part:

Dear Judy:

In order to protect the study's integrity, we are unable to support an itinerary that includes a stop in DC. I am happy to book you a direct flight from LAX

to NYC. If you will be stopping in Washington DC, unfortunately we will be
unable to host you at Columbia.

Allison M. Kanas

Project Manager

Center for Infection and Immunity

Mailman School of Public Health

Columbia University[17]

Holy crap! How does stopping in Frederick, Maryland to visit her mother
affect the integrity of any study? At the planning meeting in Lipkin's office,
Lipkin received calls from John Coffin and Harold Varmus.

Clearly, Lipkin was only the puppet master in this study who would get
his financial reward in the end.

Neither Judy nor I thought Lipkin was the originator of the insulting
gag order. Shortly thereafter, Judy received another email, this one from
a member of Anthony Faucis's staff, which read, as I recall, something to
the effect that if she entered the Frederick Cancer Research Facility at Fort
Detrick, she would be escorted off the grounds by NIH security.

Years later, on her mother's eightieth birthday in 2016, Judy went to Fort
Detrick to check on her retirement status. She was still listed as a "fugitive
from justice" who would be arrested if she set foot on the federal property for
any reason. In my opinion, this could only have been orchestrated from the
very top, by people like Harold Varmus, Anthony Fauci, and Francis Collins.

I'm sure these three will deny doing anything so odious but does anybody believe this would be done without the full knowledge of the head of
the NCI, NIAID, and the NIH?

I had already gotten an email from Harold Varmus in September of
2010, shortly after the publication of the Alter/Lo confirmation of our work.
The email read in full:

Frank:

As you have probably heard, Francis Collins, Tony Fauci, and I and a few others met recently to discuss plans for NIH's coordinated activities in response to
concerns about the confusion surrounding XMRV, polytropic MLV, and CFS
(and also prostate cancer.) Ian Lipkin is designing a study to help with this, but
the group in Bethesda is pulling together some confidential reports of work in
progress at the NIH, as well as published work, to help guide our thinking.

> We would appreciate receiving from you sometime this week a short summary of your unpublished findings that bear on this matter and your short-term plans for addressing the controversial issues. That report will not be circulated beyond the small group that will continue to meet with Francis and will be treated with the confidentiality it deserves. Thanks for your help and let me know if you have any questions.
>
> Best, Harold.[18]

Does this read to you as if an honest investigation was being planned?

Maybe the plan was to make Judy a criminal so nobody would believe anything she says and ruin her life for a vengeful scientific establishment?

Perhaps not, but no other explanation makes more sense to me.

* * *

Just as we were starting the 2012 Lipkin study, the Lo, Alter et al., paper was retracted with the following statement "although our published findings were reproducible in our laboratory and there had been no evidence of contamination . . . we feel the association of murine gamma retroviruses with CFS has not stood the test of time."[19] This statement obviously satisfied the powers that be as Alter shared the 2020 Nobel Prize in Medicine.

Meanwhile, the study design for the multicenter study, which had been decided November 4, 2010, and what was actually done were two very different things.

Participants with co-morbidities and coinfections typical of patients with ME/CFS were not included, even though sicker people are more likely to express the virus.

The VP-62 infectious molecular XMRV clone spiked samples were not supposed to be included because of the risk of containing a sample containing the defective or slower replicating natural isolates.

We were told that if we detected a positive in one of the duplicates and a negative in the other, we'd get a third sample to break the tie.

Many of the samples had the positive/negative split scenario when Fauci stopped our ability to get that third sample to test. Lipkin then stepped in and applied their "teacup statistics," which he said determined that those results were contamination and we did not need to test a third time.

Even with all these issues, there were positives in the chronic fatigue syndrome cohort, as well as in the healthy samples.

The study was reported as showing no evidence of XMRV/MLVs viruses, or association with chronic fatigue syndrome. The truth is there was evidence of infection in 6 percent of patients and in 6 percent of controls. The title of the paper actually stated there was no evidence of XMRV in humans. Despite his promise to collaborate with us on future investigations, Lipkin never contacted us again.

The rest of 2012 and 2013 did not go well for me. The NCI administration decided to launch an investigation into whether there was any fraud in my performance of the XMRV project. No fraud was found in any lab associated with any of these original papers.

The failure of other labs to reproduce data is not a reason for retraction and occurs all the time. Neither is the omission of a step/reagent (5-Azacytidine treatment of samples to activate methylated viruses in the methods) a reason for retraction.

Later, a paper showing restricted replication of XMRV in Pigtailed Macaques was published, suggesting primates could be infected.[20]

Why not humans?

On September 10, 2013, Lipkin was on a CDC conference call on the radio where he said they had done high throughput sequencing on hundreds of samples. They detected retroviral sequences in the blood or plasma of 85 percent of the samples but did not think they were going to pan out to be significant. This is important because, using a more sensitive technique, he confirmed that the blood supply was contaminated with gammaretroviral sequences, thus confirming the earlier work that his study just denied. Apparently, this public conference call on behalf of the CDC was not recorded. But I have a contemporaneous email from Dr. Maureen Hanson of Cornell University, in which she relayed what Lipkin said on September 10, 2013:

Lipkin was on a "CDC conference call." I think they recorded it and it might be available eventually though I'll bet it will take at least a week.

I was able to get on the call after the first ten minutes, as I was on a PHD qualifying exam before it. I was quite surprised at how much info he gave out. Basically, he said they had done high throughput sequencing on 100s of samples. He said they detected retroviral sequences in 85% of the samples, but I don't think he said whether these were cases or controls. He emphasized that he didn't think these were going to pan out as significant (my guess is they

> might have been HERVs? Though he didn't say that.) He also said that they
> hadn't yet found a virus or microbe that seemed to be associated with CFS.
>
> He actually was complaining about not having enough money for microbi-
> ome analysis, not for virus hunting. Wants to look in fecal samples for both
> bacteria and fungi.[21]

How is it that Ian Lipkin can find evidence of a retrovirus in 85 percent of his samples (presumably from the well-characterized cohort of Stanford researcher. Dr. Jose Montoya) and say he doesn't know if it's relevant? Can we at least agree that it's a still an open question?

In 2017, a paper reported that Italian CFS patients were 4.3 percent XMRV positive.[22]

There were persistent rumors that the Board of Scientific Counselors was going to look at my budget and cut out all virus funding. This was a not-so-subtle attempt to get me to retire by making things difficult to continue. I had no guarantee that I would pass a new lab review. This was personally devastating as we likely found a new retrovirus, which I would never be able to characterize.

* * *

On September 30, 2013, the official day of my retirement, I got a call from Robert Gallo, whom I hadn't spoken with since 1985. Gallo said, "Frank, Kendall Smith is writing articles that demeans my role in the discovery of IL-2. Please write and defend me."

I said, "Bob, I'm retiring. It's your problem now."

Gallo then said, "I'm sorry how things turned out between us."

I replied, "Bob, that could not be a weaker, more disingenuous apology and it's way too late."

By all rational measures, I have been lucky in life. I've had decent parents, a great mentor, a supportive and loving wife, great siblings, and a wonderful son. I've also had many caring and intelligent people with whom I've worked.

As with most well-off people across the political spectrum, my pride tells me I earned it. My humility tells me to recognize all the advantages I was given to enable me to come from the humble circumstances to live the life I have.

With that recognition comes the responsibility to do something to assist people who were not afforded these advantages. I think I could have done better to eliminate any vestiges of misogyny in the work place and the larger world.

The collusion between the regulators and their friends in corporate America is destroying our health.

This is where my perspective and that of Judy differs, just as it did forty years earlier on the question of Ralph Sampson and whether he should play for the Boston Celtics.

Judy believes most of their actions are deliberate.

I think they generally display and rely upon an arrogance leading to stupidity born out of the corruption of power, particularly to cover a previous mistake.

But the harm they cause, whether intentional or unintentional, is the same, and deserves our attention.

In 1995, John Coffin and Jonathon Stoye composed a letter to the editor of *Nature* about the danger of viruses being transferred from animals to humans by the transplantation of animal organs (or tissue) into human beings, a process known as xenotransplantation. They wrote:

> Throughout this episode, there has, however, been little discussion of the dangers posed by the possible activation of endogenous retroviruses. Such viruses are widely distributed in mammalian species including pigs and baboons, potential donors for these procedures. Since they are inherited in the germ line in the form of proviral DNA, they are impossible to remove using the usual methods for deriving pathogen-free animals. Many endogenous retroviruses cannot replicate in their native hosts but will grow to high titers on heterologous [different] cells, a phenomenon known as xenotropism . . .
>
>   . . . Indeed, transplanting an organ carrying endogenous xenotropic provirus is equivalent to injecting a patient with a live C-type virus.[23]

In 2021, they do not seem to be concerned at all because our immune system is so strong. But what about the people who have problems with their immune systems?

Doctor David Kessler, who ran the FDA from 1990 to 1997 when OxyContin came out on the market, now says, "The opioid epidemic is one of the great mistakes in modern medicine."[24]

Judy does not believe it was a mistake.

I wonder, *Why didn't they do their job more thoroughly?* It only took them twenty years to figure that out. And what happens to these decision-makers with blood on their hands, like David Kessler?

Kessler got another job as the chief science officer of the White House COVID-19 response team. So much for accountability in science.

All these drug companies are facing huge penalties to pay off lawsuits.

One company is allowed by the FDA to use future drug sales, including OxyContin, to pay off lawsuits.[25] That's right! Kill and harm more people to pay off the people you've already killed or harmed. The same goes for chemicals like glyphosate and the weed killer, Round-Up, or aluminum-laced talcum powder.[26]

Personally, I do not think the argument over the origin of SARS-CoV-2 is important, except that Anthony Fauci has misrepresented every other important public health issue since he was appointed director of NIAID in 1984. Public health officials send so many mixed messages, it's a wonder anybody believes them. We and our enemies have been making bioweapons long before anybody heard of Wuhan. And they are often released, through stupidity, often into the world.

We must admit that public health is a global problem. AIDS is still a pandemic in sub-Saharan Africa where it is the leading cause of death for women of child bearing age.[27] Until we have the will to support these people's basic needs—clean water, clean air, and good food—there will not be much change. The vaccines Bill Gates keeps trying to force upon the world will not help alleviate basic needs. In this country, we need to look at our Native American reservations and the poor in Appalachia, and the shame of our inner cities.

During all the XMRV fallout, I heard from an old friend, Dr. Connie Faltynek, from the Biological Response Modifiers Program and our interferon trial days. She went to work for the pharmaceutical industry to find a drug that worked for pain and had fewer of the side effects of opioids. She found that:

> Antagonists of the capsacin (active ingredient of hot chili peppers) and cannabinoid receptor TPRV (transient receptor vanilloid type 1) have efficacy in many preclinical pain models. However, the data suggested potentially profound drug induced thermosensory impairment.[28]

The drug decreased pain, but also sensitivity to heat. Judy recognized that CBD (cannabinoid) could interact with TRPV1 therapeutically and decided to work on it with compassionate producers in California. Science works better when we all work together.

I have become intrigued by the health benefits of natural products like cannabis, mushrooms, probiotic bacteria, short chain fatty acids, and bacteriophages.

It's everybody's obligation to do something with the gifts they are given. Although I am officially retired, I continue to pursue these questions.

* * *

I hope that what you've read from me lets you know what it was like for a scientist on the inside who spent thirty-eight years in government research. Unlike Judy, I have never spoken publicly about these issues. I don't know if that gives me greater credibility or less, because I have remained in the shadows for so many years.

All my adult life, I have only wanted my body of work (the data) to speak for me, but perhaps that is not enough since I've come to the realization that nothing in life is of value unless it is shared with others.

I have simply provided professional and private data for my story.

The interpretation you put on that story is up to you.

PART TWO

Judy's Perspective

All who, while unable to be saints but refusing to bow down before pestilences, strive their utmost to be healers.
                                        —Albert Camus, *The Plague*

PROLOGUE

Keep it Simple, Judy!

People often say to me, "Judy, I know you've been in science for more than forty years. But I haven't. Can you simplify your ideas for me so I can follow along?"

It's a complaint I've heard more times than I care to admit. Once, after giving a talk, a woman came up to me and said, "Dr. Mikovits, that was the best talk on any subject I have ever heard in my life."

I stood a little taller and replied, "Why, thank you. What did you like best about it?"

She cheerily replied, "Oh, I didn't understand a word of it. It's just the way you said it."

I was crushed, mortified, and thankful Frank Ruscetti had not heard it. He spent so many years trying to teach me transitions and how to make slides that everybody could understand. Yet, in 2010, in London, I had completely and utterly failed.

In 2021, I'm in a unique position, trained in speaking to other scientific experts, and yet thrust into the public limelight of the thirty-second sound bite and media attack. When your opponents can't win the argument, they censor you. Previous critics of our books said I used jargon.

Thank God in 2021, PCR is no longer considered jargon, and many people express strong opinions about the use of PCR tests for COVID-19.

I'm doing my best, sharing the data, at a time when the mainstream media doesn't want to let people like me speak, because if the public understands what we're saying, it will threaten some very powerful corporate and governmental organizations.

Besides writing books, where my wonderful coauthor, Kent Heckenlively, does his best to simplify the science and yet keep it accurate, I often give talks, such as I did in October 2019 in Anaheim, California at *The Truth About Cancer* conference. I meticulously prepared my slides, listing my claims, as well as the scientific publications, which back up what I'm saying.

But most of all, I tried to keep it simple.

I did not know it, but this conference would be like no other with respect to the impact it would have on my life. A little background is necessary. The invite came by email from my dear friends, Ty and Charlene Bollinger, in January of that year. My mother was in her last days on Earth so I wasn't paying much attention to email. I saw the invitation and quickly looked at the date of the conference, as to whether it was even possible for me to attend.

October 12–15, 2019 in Anaheim, California was perfect, just a few days after my wedding anniversary. I also appreciated the $2500 travel stipend and honorarium. I did not think about it much as my mother passed away on January 22. She requested of my siblings that the funeral be quick so I could continue my important work exposing and ending the plague of corruption in medicine that prematurely ended her life.

So, I did not think much of it as I was to participate in an ACIP (American Committee on Immunization Practices) meeting from February 27–28; a stem cell meeting March 22 –24; and had a late summer deadline to complete our book *Plague of Corruption*, before fall meetings starting with one of my favorite groups, the Health Freedom Idaho Conference from September 21–23. It was a busy time and I welcomed it to get my thoughts off of my mother's death.

I got an email reminder that my slides for *The Truth About Cancer* conference were due the next day. Not a problem, I thought. I'll just flip the slides from Idaho, and instead of discussing causes, I'll focus on cannabinoids and natural products. Then I saw the title of the talk they wanted me to deliver: **Persecution and Cover-up**.

"Not again," I thought. I wondered if I'd ever be allowed to simply move forward and enjoy my passion: formulating natural products as medicine.

Almost immediately I got a text from Frank, which read, "Dr. J., how can I help?"

Good ole Frank. I guess he saw the email and read my mind.

I texted back: "Yeah, can you find that mug shot in *Science*?"

"F**K NO!" he texted back. "I'm not ruining my weekend!"

For the first time since 2011, I was going to have to do this myself. I opened my web browser, typed in sciencemag.org, then put in my name and Jon Cohen, the writer who had apparently been tasked with the job of destroying my scientific reputation.

The first substantive slide for *The Truth About Cancer* presentation was about the two books I'd co-written with Kent, *Plague: One Scientist's Intrepid Search for the Truth About Human Retroviruses and Chronic Fatigue Syndrome (ME/CFS) and Other Diseases* (November 2014) and *Plague of Corruption: Restoring Faith in the Promise of Science* (originally scheduled to be published on November 5, 2019, but eventually published on April 14, 2020, just about a month into the nationwide COVID lockdown).

I showed a picture of both books, as well as a short quote from each book to hit the main points. From *Plague,* there was the quote, "That is the inconvenient truth. Retroviruses and environmental toxins led to this explosion of chronic diseases."

From *Plague of Corruption* the opening quote came from the 1953 movie, *Martin Luther.* It read: "To go against conscience is neither right nor safe. Therefore, I cannot and will not recant. Here I stand. I can do no other. God help me, amen."

I guess you could say this is a story of science and faith. I'm trained as a scientist, having worked in some of the most terrifying places on Earth, such as the Biosafety Level 4 lab at Fort Detrick, Maryland on behalf of the United States Army Medical Research Institute of Infectious Diseases, investigating such diseases such as HIV and Ebola. And yet, I'm also a person of deep religious faith. I do not store my riches in this world.

Frank used to joke with my husband David about my spending on cannabis for my friends in need. He'd say, "David, if I won the lottery tomorrow, how long would it take your wife to give it away?"

Not missing a beat, David would say, "You don't understand, Frank. She already has."

I strongly believe what I do in my time on Earth is of eternal consequence.

Many times, people have asked me, "Judy, can't you just tell a little lie and make things easier on yourself?" The simple fact is I can't because the Ten Commandments are not suggestions; they are commands of the Lord God. And I fear only the Lord.

The second slide was from a senior member of the Phoenix Rising website (a site dedicated to chronic fatigue syndrome/ME/CFS), named CBS, commenting on February 23, 2011, about how threatening the XMRV story was to the public health agencies:

> "Agency heads are scared to death of how the patient population will react if XMRV works out." – Suzanne Vernon, September 11th. Lobby of the Salt Lake City Downtown Hilton – During a break at the 2010 Offer Utah Patient Education Conference.
>
> I've been struggling with what I ought to do with this for almost six months. Suzanne Vernon said this during a conversation she was having with me and Cort [Johnson]. She just sort of interjected it. No real need, nor was there much of a segue. She said that it should not be repeated. Yet I wondered why on Earth she would say something like that to someone she had just met.[1]

Now, who is Dr. Suzanne Vernon? At the time she was the scientific director of the ME/CFS Initiative (formerly the CFIDS Association of America), and from 1990 to 2007 she'd worked at the CDC, leading the Chronic Fatigue Research for ten years.[2] If there was anybody who would know the thinking of agencies regarding chronic fatigue syndrome/myalgic encephalomyelitis and the patient community, it would be Suzanne Vernon. Perhaps her conscience got the better of her for a moment and she was briefly honest about the danger of my work to institutions that had apparently been asleep at the switch for decades.

But I doubt it.

Note the date of her alleged comment. September 11, 2011 was only a few days after I'd presented my talk showing high levels of XMRV in a cohort from London and 4 percent in the controls my collaborator obtained from the blood supply in London. It was that talk which, in my opinion, prompted Francis Collins, head of the NIH, to direct Anthony Fauci's NIAID to fund a large multi-center study of the prevalence of XMRV in patients with chronic fatigue syndrome in the United States, to be headed up by Dr. W. Ian Lipkin of Columbia University.

And there's additional support for the idea that my work was a danger to the public health institutions as detailed in a 2013 article in *Discover* magazine, written by Hilary Johnson. She detailed how on July 22, 2009 I'd been part of an emergency one-day conference at the NIH, several months before our paper would be published in the journal *Science* in October 2009.

> At the center of the speculation about the new retrovirus that day in July was an immunologist and AIDS researcher named Judy Anne Mikovits (pronounced

My-ko-vitz), a diminutive 51-year-old in a sleek black suit and a crisp white shirt. Mikovits was a 20-year veteran of the National Cancer Institute (NCI) who had coauthored more than forty scientific papers. During her final two years at the agency, she had directed the lab of Antiviral Drug Mechanisms, where she studied therapies for AIDS as well as one of its associated cancers, Kaposi's sarcoma . . .

Given such far-reaching implications, it was not surprising that when Mikovits stopped talking, nearly a minute passed before someone spoke, and then it was to say, "Oh my God."

That simple expression of dread was the preliminary gasp in what would become, in the three years that followed, the most clamorous scramble in recent medical history to prove or disprove what seemed to be a viable hypothesis, one so dire it was facetiously dubbed the 'Doomsday Scenario,' by one skeptic.[3]

There it is in black and white in one of the most prestigious publications of science for the public.

I was putting forth a "doomsday scenario" which had gone unnoticed by the experts in public health for decades.

The third slide posed the question of what the agency heads would do in light of this information which strongly suggested they'd been negligent, if not worse, in the previous decades. They would, in succession, get the authors to destroy the data, force the authors to withdraw the data, then when Frank and I wouldn't comply with their directive, they force retracted the paper, without our approval. This was in spite of the fact that our research was confirmed by future 2020 Nobel Prize winner, Harvey Alter, his coauthor, Shyh-Ching Lo, and others in September 2010. This is from their publication in the *Proceedings of the National Academy of Sciences*:

Our laboratory detected MLV-related virus *gag* gene sequences in DNA from PBMC [peripheral blood mononuclear cells] and whole blood samples from 32 of 37 (86.5%) CFS patients compared with 3 of 44 (6.8%) volunteer blood donors, using a two-round nested PCR. Following only one round of PCR amplification, 21 of the 41 CFS patients' samples were found positive compared with only 1 of 44 donor samples. In every instance throughout these studies, the "positive" results by PCR (an amplicon of the predicted size) was confirmed by sequencing.

In four CFS patients from who two samples were obtained two years apart, the *gag* gene sequences were detected on both occasions. Further, *gag* gene sequences were still detectable in seven of eight CFS patients from whom

fresh samples were obtained, fifteen years after they were initially found to be
MLV *gag* gene positive.[4]

Lo and Alter's study found a greater diversity of mouse (murine) leukemia
viruses in chronic fatigue syndrome patients, but their results were even more
robust than our findings. They found evidence of the virus in 86.5 percent of
the patients, and 6.8 percent of the controls. Extrapolated to our current US
population of about 328 million people, that meant more than twenty-two
million of our fellow citizens were carrying this virus around like a ticking
time bomb, just waiting for some immune system challenge to detonate it.

And of course, they were forced to issue a modified retraction of the
paper. In essence, the retraction was: "We know this blood was from the
late 1980s and late 1990s, but since we don't have a time machine to go back
and get more blood from these same patients, the results are questionable."

Really, I encourage you to read the supposed "retraction" and see if you
come away with a different explanation.

* * *

The fifth slide was titled "Who and Why?" and showed a picture taken
in 2011 in front of the NIH Clinical Center. In the center of the picture
was Hillary Clinton, flanked on one side by Anthony Fauci and Francis
Collins, while on the other side was Harold Varmus, along with two other
scientists.

Next to their picture I had the program from the July 22, 2009 workshop entitled "Public Health Implications of XMRV Infection." Are you
scratching your head, saying, "Judy, I thought your XMRV paper was
released in October 2009? That's two and a half months before you released
your findings."

You would be right.

Our research didn't take the government by surprise. They were already
working on an answer. My biggest mistake? Trusting the government to do
the right thing.

The organizers were Stuart Le Grice (HIV Drug Resistance Program
and head of the Center of Excellence in HIV/AIDS & Cancer Virology) and
John Coffin (Tufts University and office of the Director, Center for Cancer
Research).

In addition to me, the participants were: Carlos Gordon-Cardo
(Columbia University), Stephen Goff (Columbia University), Eric Klein

(Cleveland Clinic), Robert Silverman (Cleveland Clinic), A. Dusty Miller (Fred Hutcheson Cancer Research Center, University of Utah), Stephen Hughes (HIV Drug Resistance Program, NCI), Vineet Kewal-Ramani (HIV Drug Resistance, NCI), Douglas Lowy (Laboratory of Cellular Oncology, NCI), John Schiller (Laboratory of Cellular Oncology, NCI), Chris Buck (Laboratory of Cellular Oncology, NCI), William Dahut (Medical Oncology Branch, NCI), James Gulley (Laboratory of Tumor Immunology and Biology, NCI), Jeffrey Schlom (Laboratory of Cellular Oncology, NCI), W. Marston Linehan (Urologic Oncology Branch, NCI), and Charles Rabkin (Division of Cancer Epidemiology and Genetics, NCI).

Besides me, there were eighteen other participants at that meeting, representing Columbia University, Tufts University, the Cleveland Clinic, and nine representatives of the National Cancer Institute.

Our publication on October 2009 didn't catch any of them by surprise.

In fact, they delayed the publication as long as they could. I believe this because our samples and data had been shared with all who were at the July 22, 2009 meeting and had requested them. All had confirmed our results by September. We began to hear rumors of the pending publication and became concerned because a breach of confidentiality is grounds for *Science* NOT to publish a paper.

Those were times of many sleepless nights.

The sixth slide was entitled "21st Century Acquired Endocannabinoid Immune Dysfunction: Unintended? Consequences of Unsafe Vaccinations and CDC Schedule." I listed many types of cancers as being associated with an endocannabinoid deficiency, as well as diabetes, cardiovascular disease, Gulf War Syndrome, autism, chronic fatigue syndrome (ME/CFS), and several psychiatric disorders such as obsessive-compulsive disorder and post-traumatic stress disorder. Critically, I put an asterisk next to every disease where at least one peer-reviewed publication of a retroviral association was suggested for the disease.

The seventh slide was entitled "How Did Mouse Retroviruses Get into Humans?" I'd struggled to explain the concept of using animal tissues to grow viruses for vaccines and how that might transfer an animal virus directly into the human bloodstream, when my good friend, the radio host, Ernie Hancock, said, "So, basically it's like one of the Vego-Matics, where instead of chopping up vegetables, they're chopping up animal parts, processing them, then injecting them into people?"

(This is certainly not the only way these viruses can get into people. Our work at that time was completely focused on the blood supply because of our experience with HIV and the many biological therapies derived from blood products.)

I replied, "Exactly, that's how polio, MMR, chicken pox, and the shingles vaccines are made."

Then a cartoonist friend turned it into a visual, with a conveyor belt of mice leading them into a "Mouse-O-Matic," which chops them up, then a hose siphons the red fluid into another tank labeled "Vaccines," where it's processed, and from there a nurse draws out some of the material in a syringe to inject into a baby.

The cartoon was a crude simplification, but even others picked up on the likelihood that this mouse virus had been transferred to humans in the common practice of using animal cell lines to grow viruses for vaccines. An article in the January 2011 issue of *Frontiers in Microbiology* described it this way:

> One of the most widely distributed biological products that frequently involved mice or mouse tissue, at least up until recent years, are vaccines, especially vaccines against viruses . . . It is possible that XMRV particles were present in virus stocks cultured in mice or mouse cells for vaccine production and transferred to the human population by vaccination.[5]

It's easy to understand how the first researchers developing vaccines wouldn't have had any inkling about the various viruses residing in these animals. But as the technology developed over the years, did anybody stop to check old assumptions? I'm sure it must have occurred to somebody to use the latest technology to see if our current vaccines were contaminated with other viruses.

But who wanted to be the first to step over that line and challenge one of the main pillars of public health, in addition to a pharmaceutical industry which makes billions of dollars selling that very same product?

* * *

People always ask me, "Judy, why were you arrested, held in jail for five days without bail, your notebooks stolen from you, prevented from posting on social media for five years, then when you tried to bring a civil action against those who had persecuted you, the case was held under seal for five years by the state of Nevada, and apparently, the federal government, as well?"

Slides six, seven, eight, and nine showed answers with evidence for each of those questions.

Slide six showed an email I sent to Simone Glynn at the NIH on August 31, 2011, at 8:24 p.m. PST, regarding the rush to publish the results of an investigation into XMRVs in the blood supply.

But let's be accurate here.

The study was NOT designed to determine whether the blood supply was contaminated, but whether we could develop a quick diagnostic test. I believe it was that email which precipitated the final decision to try and destroy me, both professionally and personally. Here are some excerpts from that email:

> . . . Given the complexities and limitations of this study, many of which were not recognized at the time, the (flawed) experimental design was agreed upon, to have one day to agree on a manuscript, a holiday at that, is totally unacceptable. This is NOT good science or the appropriate process. What is the rush?
>
> Afraid [of] the truth? How many of these viruses were introduced into the human population and are now threatening a lot more than the blood supply??! Because a few declared it "impossible" 40 years ago and JC [John Coffin] himself was the most vociferous!
>
> How many XMRVs?
>
> I am sending this only to Simone and Frank *because I will make this entire rush a public relations nightmare for the entire US govt.* (bold and italics added.)
>
> I have integration data and variants of many new strains!! Did those arrogant SOBs introduce these into humans and are now trying to cover it up?
>
> And then pedigree the negatives with a test with a cutoff so high it would not find a willing Roman in a whorehouse?
>
> Wonder if anyone will listen to a press conference from me?? Asking how many new recombinants from vaccines? From lab workers? Doctors? The first ever contagious human retrovirus???? Spread like mycoplasma?? Are you kidding me???
>
> It happened once!!! How many xenograft lines were created?? How many vaccines contained mouse tissue??
>
> These sick people lost their entire lives and this travesty of justice will not be carried out at their expense. Not again . . .

Nothing about these data say anything about Lombardi et al [our paper] or Lo et al [the confirmation by 2020 Nobel Prize winner Harvey Alter and his team]. Except that there are likely to be many strains of XMRVs and only God knows the impact on chronic disease. But nothing about this study says anything about our original discoveries.

And if this is rushed to print without a fair and balanced discussion of its limitations, I will spend every minute of my life exposing the fraud that has been perpetrated against this patient population.

—Judy Mikovits[6]

Yeah, I've been known to get a little heated when it comes to the health of people and scientists not telling the truth. In retrospect, I'm sure I sounded like the Gerard Butler character in *Law Abiding Citizen* (2009), who declares at one point, "I'm gonna pull the whole thing down. I'm gonna bring the whole fuckin' diseased corrupt temple down on your head. It's gonna be biblical."

But just as there was a problem, there was a solution to the blood supply issue on the horizon. It was a blood decontamination technology called INTERCEPT, created by the Cerus Company of Concord, California.

Slide seven talked about the problem with the blood supply, as well as the solution. On March 29, 2011, I gave a talk at the New York Academy of Sciences, noting that all the various studies had detected anywhere from 4–8 percent of the healthy population had antibodies to XMRV, and were thus at risk. In that talk I listed my five conclusions:

1. Data suggest there are different strains of Gamma Retroviruses that can infect humans.
2. Assays that capture the variation of these viruses in the blood supply are the best, i.e Serology and transmission.
3. Cerus Technologies can inactivate infectious strains of XMRV/HGRVs in blood components.
4. New disease associations include leukemia, lymphoma, and the platelet/megakaryocyte disorder, ITP.
5. We need more full-length sequencing.[7]

On the slide, I noted that the FDA had approved the INTERCEPT system on December 1, 2014, as well as an article in *Scientific American* on June 16, 2015 with the title, "The INTERCEPT Blood System Rids Donations of All Pathogens."[8]

You see, I'm in agreement with *Scientific American* that the Cerus INTERCEPT system works. But my position is logical.

The blood supply is contaminated and needs to be decontaminated.

The public health authorities are saying the blood supply is fine but still needs to be decontaminated. In July 2020, Wikipedia was reporting the blood supply was being tested for XMRV? Why, if they cleaned it up in 2012 with the Cerus INTERCEPT system?

Slide eight showed a quote from what Frank told me John Coffin said to him in November 2010, when XMRV was still allowed to be a viable theory. Coffin said, "*Science* started this, and *Science* is going to end this!"

In other words, the journal *Science* was going to be our judge, jury, and executioner.

The study on the blood supply was rushed into print without my approval on September 22, 2011. Remember, this was supposed to be a study about designing a test to detect the virus in the blood supply. Not whether the virus was in the blood supply. That had already been confirmed by various studies.

A week later on September 30, 2011, Jon Cohen wrote the hit piece for *Science*, with the title "The Waning Conflict over XMRV and Chronic Fatigue Syndrome." Here's the opening of the article:

> OTTAWA, CANADA - Less than a day after a new study dealt what many consider a lethal blow to the controversial theory that a newly detected virus, XMRV, is linked to chronic fatigue syndrome (CFS), proponents and skeptics of the theory squared off in a meeting here.
>
> In one corner was Judy Mikovits, research director of the Whittemore-Peterson Institute for Neuro-Immune Diseases (WPI) in Reno, Nevada, and the main champion of the idea that XMRV and its relatives play a role in CFS. Her opponent, an erstwhile supporter, was heavyweight retrovirologist John Coffin of Tufts University Sackler School of Graduate Biomedical Sciences in Boston.[9]

Funny how my main opponent in this debate, John Coffin, was the same guy I'd named in my email, who forty years earlier had told people not to worry about how animal viruses in tissue cultures used for vaccine production might infect people, and the blood supply as well.

How did Jon Cohen know that on September 29, 2011, I'd been fired from my job at the Whittemore-Peterson Institute? Were my former employers, Annette and Harvey Whittemore, and former colleague Vince Lombardi promised to be forgiven of their crimes, which included misappropriation of federal funds and Medicare fraud (although Harvey was sentenced to two years in federal prison for campaign bribery charges involving US Senate Majority Leader Harry Reid[10]), if they put me into legal jeopardy?

How did Cohen of *Science Insider* know I'd be fired a day before the article came out, and thus, unable to respond with those all-important genetic sequences of the XMRV virus which were in the desk drawer Harvey Whittemore named in his affidavit in his original lawsuit against me? Funny how the lawsuit was refiled with a new affidavit, now of a crime from my student Max, who caught them in the middle of stealing my work in the early morning hours of September 30, 2011, thus spoiling their perfect crime?

It got worse.

On November 18, 2011, I was arrested in my own home in California for being a "fugitive from justice." How can you be a fugitive in your own home? Here's how *Science* reported it:

> Sheriffs in Ventura County, California arrested Mikovits on felony charges that she is a fugitive from justice. She is being held at the Todd Road Jail in Santa Paula without bail. But *ScienceInsider* could obtain only sketchy details about the specific charges against her. The Ventura County Sheriff's office told *ScienceInsider* that it had no available details about the charges and was acting upon a warrant issued by Washoe County in Nevada. A spokesperson for the Washoe County Sheriff's Office told *ScienceInsider* that it did not issue the warrant, nor did the Reno or Sparks police department. He said it could be from one of several federal agencies in Washoe County.[11]

Let's go over a couple of the claims in the article, which don't seem to make any sense. They say the truth is often found in the anomalies.

First, Ventura County officials said the warrant was issued by Washoe County.

Second, no warrant was issued by the Washoe County Sheriff's Office, or the Reno or Sparks police department.

Third, the only other possibility is "it could be from one of several federal agencies in Washoe County." Remember those federal agencies who were "scared to death if XMRV worked out?" I believe XMRV in the blood supply is going to be a repeat of the disaster of HIV in the blood supply.

I'd really like to know how all of this happened. To this day, nobody has given me an explanation of what happened. In an investigation for a documentary about my life, the producers said the best their investigators could come up with was that a "verbal arrest warrant" was issued for my arrest. No lawyer I've ever consulted has been able to give me any details about this so-called "verbal arrest warrant."

Perhaps the answer can be found in a later article from *Science* on February 8, 2012, with the title "Embattled Institute Retains Major Grant to Study Chronic Fatigue Syndrome."

> WPI [Whittemore Peterson Institute], based in Nevada, could have lost the grant from the National Institute of Allergies and Infectious Diseases (NIAID) because in September it fired Judy Mikovits, the principal investigator on the award. WPI subsequently filed a lawsuit against Mikovits for allegedly misappropriating property, and she also became subject to a related criminal case that led to her arrest and brief jailing. Mikovits has maintained her innocence and both cases are in court.
>
> Harvey Whittemore and his wife, Annette, who founded WPI, are defendants in a lawsuit filed against them by his former business partners who allege that the couple inappropriately used funds from a holding company he co-owned to support their institute.[12]

I'm trying to remember, who runs NIAID, and has since 1984?

Oh, that's right, Dr. Anthony Fauci.

I was the principal investigator on that grant, which means I should have been able to take that money with me to my next institution to do the work. But Tony Fauci gave it to Harvey and Annette Whittemore, who'd just been sued by their former business partners for forty-four million dollars and who had misappropriated that money.[13]

This slide is packed with defamatory statements and lies after this "beleaguered institute" was repaid for silencing me with millions in grant money. Jon Cohen's partner in crime, Martin Enserink, also wrote defamatory articles about me for *Science Insider*, saying I had a "less than flattering reputation." The next slide added discussion of fraudulent publications by Coffin, Knox, and Peterson earlier that year, which we successfully defended to the journal editors, arguing that ONLY the Silverman PCR should be retracted from the paper.

Since Silverman knew he had committed fraud in the original paper (not revealing that his VP-62 molecular clone of XMRV was in fact created from three different patient samples, rather than a true isolate), stating, "I requested a full retraction of our findings this summer after discovering that the blood samples were contaminated, implying they were contaminated in our laboratory. I was in favor of a retraction of the entire paper, and I agree with the decision." Frank Ruscetti has emails from Silverman and his apology at the meeting in Leuven, Belgium, June 6–9, 2011, where he told Frank, "Joy [Jaydip Das Gupta] is really sorry about what happened." Was Silverman sorry that his laboratory had deliberately contaminated the samples with VP-62? Of course, Silverman's emails to this effect will be happily provided in the event of any litigation.

Science then tried to land the final blow to end my career in December of that year when they retracted our original 2009 paper in an article entitled "In a Rare Move, *Science*, Without Authors' Consent Retracts Paper that Tied Mouse Virus to Chronic Fatigue Syndrome."[14] Of course, it didn't hurt that I'd just gotten out of jail. Science could show that mug shot and send a clear warning to anyone who dared to go against Fauci and the misogynistic gatekeepers of science again.

But we weren't supposed to worry that *Science* had forced a retraction of our paper. We had the Ian Lipkin multi-center study to determine the ultimate truth of our findings. As reported by the *New York Times* on November 22, 2010:

> Dr. Lipkin is now preparing to dive into a new controversy. In October of 2009, a team of scientists claimed to find a virus known as XMRV in people with chronic fatigue syndrome . . .
>
> In September, Dr. Fauci asked Dr. Lipkin to organize a large-scale investigation. Earlier this month, Dr. Lipkin brought together scientists from three labs that have gotten contradictory results to being working together on a new search for the link.
>
> "There isn't anybody better than Ian Lipkin," said Dr. Fauci. "If he can't find it, it probably doesn't exist."[15]

When we published our first book in 2014 and mentioned Fauci's name as the dishonest architect behind the XMRV investigation of Ian Lipkin, few people had heard the name.

The situation is much different in 2021 as he is probably the most divisive figure in America today.

* * *

Let's talk about that Lipkin study and three of their conditions which were excluded.

The Lipkin study disqualified anybody with a "medical or psychiatric condition that might be associated with fatigue," "abnormal serum characteristics," or "abnormal thyroid functions."[16]

I remind people the condition they were investigating is called chronic fatigue syndrome, but they were excluding anybody with a "medical or psychiatric condition that might be associated with fatigue." How is this science? Well, as we know from the presentation title, this was never about science. It was about making XMRV go away.

And if you understand the mindset of these people, you realize they will say the truth quietly, preparing themselves for that moment when the wind shifts. The cohort I was most excited about in the Lipkin study was that of Stanford scientist, Jose Montoya, whose patients actually had the immunological signature of disease as in the original study.

But of course, the study was stopped by Anthony Fauci before that cohort could be tested.

However, about a year later Ian Lipkin reported the results of the Montoya cohort in a public conference call with the CDC on September 10, 2013.

> We found retroviruses in 85 percent of the sample pools. Again, it is very difficult to know whether or not this is clinically significant or not. And given the previous experience with retroviruses in chronic fatigue, I am going to be very clear in telling you, although I am reporting them in Professor Montoya's samples, neither he, nor we, have concluded that there is a relationship to disease.[17]

Do you understand how corrupt these efforts are? The Lipkin study supposedly demolishes the XMRV retrovirus theory and then a year later quietly confirms the retrovirus theory.

A study by an Italian lab of the MMRV (Priorix Tetra) vaccine also showed that there were retroviruses in commercial vaccines.

It was possible to confirm the presence of contaminating viruses in the following:

- Human endogenous retrovirus K – 32 sequences
- Equine infectious anemia virus – 2 sequences
- Avian leucosis virus – 2 sequences
- HERV – H/env62 – 4 sequences

These viruses are known to be adventitious vaccine contaminants and are
known to be potentially dangerous, which is why manufacturers are required
to verify that they are completely absent from the vaccine.[18]

How can we call it science when we don't use the latest technology to test
what's in our vaccines?

And as I've shown with data since the early 1990s, the so-called
"defective viruses" where you might only have a partial sequence can
still cause disease. When we use animal tissue or aborted human fetal
tissue to create vaccines, biologics, or xenograft procedures (animal tissue
being placed in human beings), there is a danger of transmission of these
viruses. The next slide discussed the study from Gary Owens' laboratory
at the University of Virginia. The paper concluded: "ENV proteins from
both viruses impact tumor pathogenesis."[19] That is, changes in microvasculature similar to vascular pathologies seen in ME/CFS and vaccine
injuries. These microvasculature aberrations are caused solely by XMRV
ENV (envelope) protein.

From the Owens/Coffin lab publication in 2013:

> Although it is highly unlikely that either XMRV, VP62 or B4Rv them-
> selves infect humans and are pathogenic, the results suggest that xenograft
> approaches commonly used in these studies of human cancer promote the
> evolution of novel retroviruses with pathogenic properties. Similar retrovi-
> ruses may have evolved to infect humans.[20]

I spent some more time on the slides explaining xenotransplantation and the
abundant concerns scientists have had about this risk of transmission. But
vaccines and injected biologics are xenotransplantation and have the same
risk of these animal viruses being transferred to humans.

Is this the same John Coffin who in the 1990s exposed the risk of xenotransplantation, but now in 2021 says that the exact same problem, this
time brought to us through vaccines, is suddenly not a problem because our
wonderful immune systems can magically handle it?

* * *

In the final series of slides, I made the following suggestions for how we can
fix these problems.

First, repeal the 1986 National Vaccine Injury Compensation Act, which gave complete financial immunity to the makers of vaccines for harm or death to children.

Second, enact an immediate moratorium on ALL vaccines until the entire schedule is tested for safety.

Third, end all vaccine mandates and restore liability to pharmaceutical companies for injury or death caused by vaccines.

Fourth, investigate and convict all criminals at the CDC, the FDA, and the NIH for their crimes against humanity.

Fifth, eliminate the Advisory Committee on Immunization Practices (ACIP).

Sixth, use the patent royalties paid to the NIH, CDC, and the FDA to compensate all the victims of this thirty-five-year plague of corruption.

This was my mindset in October 2019 about the problems in our health-care system.

I thought I was fighting for groups forgotten by mainstream medicine, chronic fatigue syndrome (ME/CFS) sufferers, autism families, and cancer patients.

As the COVID-19 darkness spread around the globe in 2020, I realized I was terribly mistaken.

In reality, I was fighting against something far more sinister.

CHAPTER ONE

Running My Own Lab without Sleeping with the Boss or Being a Lesbian

My dear readers should know that throughout my life I've often been underestimated.

But I have always tried to outthink and outwork my enemies. Maybe it's in the blood. After all, my mother's heritage was Native American (Cherokee), and my father's, Austro-Hungarian. If I took one of those popular genetics tests it might show I was related to Attila the Hun, the fifth century Germanic king whose mounted horsemen raided from the Black Sea to the Mediterranean. The Roman Empire trembled under his wrath and nicknamed him *Flagellum Dei*, which translates as the "scourge of God."

Kent Heckenlively claims descent (according to family stories) from Martin Luther, the sixteenth century religious figure who led the first successful intellectual revolt against the thousand-year domination of the Catholic Church. Luther's provocative questions laid bare the hypocrisy of the church and its leaders, with many claiming he made possible both the Reformation and the Enlightenment. Luther was a great man, but by most accounts, a good one as well. He enjoyed sparkling conversations, loved jokes, and exhibited a respect for women that seems almost modern and not from the Middle Ages. Because of his advanced degrees, his wife, Katharina Von Bora, a former nun, always referred to him as "Herr

Doktor." He returned the compliment by often referring to her as the "Boss of Zulsdorf," after the farm they lived on, or the "morning star of Wittenberg," because of her habit of rising at around 4 a.m. to begin her chores. Kent shares all these traits, from Luther's fierceness, to his compassion, to his easy laughter.

My first day running the world-renowned Lab of Anti-Viral Drug Mechanisms at the National Cancer Institute was January 4, 1999. I include here the letter sent out by Dr. Joseph Kates, director of the Research Support Programs, announcing my promotion, and giving a brief review of my history.

> I am very pleased to announce the appointment of Dr. Judy Mikovits to the position of Acting Head of the Laboratory of Anti-Viral Drug Mechanisms, Developmental Therapeutics Program, effective January 4, 1999. Prior to this appointment, Dr. Mikovits has been a scientist associated with the Intramural Research Support Program in the Laboratory of Leukocyte Biology since 1993. Prior to that, she was a post-doctoral fellow and research associate in that Program. Her research focused on mechanisms of immunopathogenesis of the human retroviruses HTLV-1 and HIV-1.
>
> Dr. Mikovits is an accomplished virologist with a number of important publications in the retroviral field. She has been interested in the interaction of HIV with the immune system and the effect of interleukins and cytokines on viral replication.
>
> As you may know, the Laboratory of Anti-Viral Drug Mechanisms was previously headed by Dr. William Rice, who left SAIC [Science Applications International Corporation, one of the private labs set up to take advantage of government contracts] to pursue antiviral drug development in the private sector. This laboratory has developed an international reputation determining mechanisms of action of antiviral drugs, particularly in the area of HIV, and working in close collaboration with other laboratories at Frederick. We are confident that Dr. Mikovits will continue this tradition of excellence as she rebuilds this Program.[1]

That's the public relations, sanitized version of my career as a woman in science up until that time. Everything he said was true, but it doesn't really tell the whole story.

Let me give you the more complete version.

* * *

Frank has already described at length the prejudice against women that he observed in the 1970s and early 1980s, both in academia and government research.

But it will probably be surprising to the average reader how many of these barriers for entry of women existed in 1987 when I began my doctoral studies at George Washington University.

When I went to graduate school, it was common that technicians in the lab would get their employer to pay for their tuition. As you went to school, you were expected to continue working in that lab. If you were a man, you could spend part of your time in the lab working on research which went toward your doctoral thesis. Between commuting and working a full-time job that meant you could expect to be busy for sixteen hours a day.

However, if you were a woman, you had to declare your doctoral thesis in advance and you COULD NOT work on those projects during your expected eight hours at the lab. You had to find additional time during the day to work on your own research, which probably meant an additional five to six hours of work, meaning you might only get about two or three hours of sleep a night.

In 1987, the data were that only one in ten thousand CD_4 T-cells were infected by HIV and yet all these cells were dying. The medical community called this "bystander effects," a term used when the science does not know what is happening, like non-infectious inflammation that was called "sterile inflammation" until it was found to be caused by dysregulation of the microbiome. We had been working with Candace Pert and Mike Ruff on Peptide T, and I knew from their work on AIDS dementia, and that of Howard Gendelman, who would be on my thesis committee, that brain macrophages are involved in the disease. We hoped that blocking the interaction of the monocyte and T-cells could prevent the development of AIDS. I declared my doctoral thesis would be concerning HIV infection of monocyte/macrophages. When doing my technician work in Frank's lab, I couldn't be working on anything to do with that, but instead focused on research pertaining to TGF-beta, stem cells, and other assignments.

During graduate school, this was my typical day: I'd arrive at Frank's lab at around four or five in the morning, put in a few hours of work, and then drive an hour away to George Washington University for my classes. Then I'd return to Frank's lab, put in several more hours on non-doctoral research, and then start my own experiments, often working until midnight or one in the morning. To put it bluntly, the men had no life, but at least they got eight hours of sleep a night.

My molecular biology professor at George Washington, Dr. Ajit Kumar, a sweet, sweet man, literally cleared a bench for me in his lab so I could perform some of my lab work at the university. I was exhausted from literally working twenty-two hours a day.

My doctoral thesis defense was November 14, 1991. My thesis was on HIV latency in monocytes and how the early administration of antiretroviral therapy could silence expression of the virus and those infected could be expected to live a normal life span if they followed this protocol. In fact, the well-known basketball player, Magic Johnson, had recently tested seropositive for HIV, meaning he had an antibody to the virus. This was front page news a week earlier. The main question I was asked during my thesis defense was, based on my research, would Magic Johnson die of AIDS?

I showed my data that demonstrated if he started the antiretroviral therapy early, maybe Peptide T or interferon alpha, and kept the inflammatory cytokines suppressed, keeping the virus latent in the monocyte macrophage, and thus prevent death of the CD4 T-cells, Magic would never develop AIDS, much less die of it. Nearly thirty years later, Magic Johnson is still with us, something that was unimaginable during the darkest days of the HIV/AIDS epidemic, when so many of these beautiful men were dying slow, horrible deaths.

In 1992, Frank received the Distinguished Service Award from the NIH. It was a well-earned accomplishment and gave both of us the feeling that good science, as well as those who actually did the work, would eventually be acknowledged.

My degree was awarded on February 17, 1992. After my successful thesis defense, Frank and I agreed it was best for me to do my post-doctoral work in the lab of Dave Derse. Derse was an accomplished molecular virologist working in the Laboratory of Genomic Diversity headed by Dr. Steve O'Brien, in a program in which they studied viral diversity in animals, with a specialty in big cats.

Since all my graduate training was in cellular biology, I wanted to develop molecular virology skills. My main project at Derse's lab was to create the first infectious molecular clone of HTLV-1. You can study a virus by producing great quantities of it, such as we did in fermentation chemistry, purifying HTLV-1 from human T-cells grown with IL-2. But that was a very labor-intensive process. Producing a molecular clone made such research a great deal easier. The problem with HTLV-1 is that it kept "popping out of the vector." I was working on that project with a researcher from Bulgaria, a brilliant woman who spoke her mind, Dr. Maria Polianova. She always made me and Frank laugh because she knew little English when she came to the Derse lab in 1992 and, with her Eastern European accent, she always said she was working as a "post-dog."

That was a pretty accurate description of how I felt at the time.

In addition to my "post-dog" work in Derse's lab, I also had to continue my work in Frank's lab. That meant my workload had shrunk from twenty-two hours a day to a scientific slacker pace of just sixteen hours. At least I was getting a good night's sleep.

A project which had grown out of my PhD thesis research was understanding DNA methylation, a main mechanism of silencing the expression of retroviruses. This work started as a collaboration with Dr. Stephen B. Baylin, a pioneer in the field of Epigenetics from Johns Hopkins University, only an hour drive from Fort Detrick. Although I was working long hours, I was basically commuting between Building 560 where the Biosafety Level 3 lab was located, and Frank's lab located in Building 567. My PhD thesis, looking at the latency of viruses, suggested retroviruses could alter the DNA methylation machinery in order to persist. How the immune system might silence the expression of viruses had generated several intriguing avenues of research, but my personal favorite was DNA methylation. That's mainly because I'm an identical twin. So, any differences between my sister and I are not due to genetics, but to epigenetics. Stephen Baylin reminded me of Frank, a brilliant thinker who could see patterns in the data and synthesize them into a new and revolutionary understanding.

At George Washington University, I did all my work using blood from HIV-infected men who had not yet developed AIDS. I isolated the various blood cell subsets and the signaling molecules to distinguish the differences in the molecules and pathways between those who had not developed AIDS and those who had developed the disease.

In addition to DNA methylation mechanisms of silencing the virus, my work showed inflammatory cytokines in patients with AIDS, but not in those HIV-infected men who had not developed the disease.

What were the key off and on signals?

I purified the nuclei of peripheral blood mononuclear cells from some HIV-infected men and showed that nuclear factor kappa B (NF-kB) was prominent in the nuclei of HIV-infected men who had developed AIDS. Nuclear factor kappa B (NF-kB or NF-Kappa-B) turned on the various proinflammatory cytokines, IL-6, IL-1, and TNF-Alpha, driving the development of AIDS. It wasn't the infection. It was, in part, NF-Kappa-B, which was causing the cytokine storm that was leading to the deadly ravages of AIDS.

We submitted that paper to a prestigious journal in 1992, and it was rejected. One reviewer wrote that everybody knows NF-Kappa-B is a cytoplasmic protein, so it was obvious my nuclei were contaminated with cytoplasmic protein, even though there was no evidence of such contamination. I had been down that contamination road before, with the earlier publication of HIV infection of monocytes/macrophages. In that paper, the reviewer rejected it, suggesting my monocyte cultures were contaminated with T-cells because everybody knew the HIV receptor is CD4 and monocytes don't express CD4. Of course, later CCR5 was shown to be a coreceptor and people who had a thirty-two base pair deletion in CCR5 would not develop AIDS no matter how much HIV was in their blood.

Frank taught me a valuable lesson as we filed the manuscript away, but not the knowledge of the data. Frank instructed me to ignore the arrogance of the gatekeepers of the scientific literature, and always show all the data, as the best papers should raise more questions than they answer. Other investigators may see clues to mysteries in our data from experiments they've done, perhaps using technology or methods unavailable to us at the time and solve the mystery.

As government researchers, we didn't have to "publish or perish" like many academic researchers at universities. However, to get a PhD at George Washington University, one needed two peer-reviewed publications. Fortunately, those two were infection of a monocyte cell line and proof of primary human monocyte infection, published a year before my thesis defense.

In the case of the DNA methylation and NF-Kappa-B papers, they took much longer to publish.

About six years later a researcher at Harvard published in *Cell*, the cloning of I-Kappa-B, an inhibitor of NF-Kappa-B. I-Kappa-B held NF-Kappa-B in the cytoplasm until phosphorylation changed the shape of the inhibitor, releasing NF-Kappa-B into the nucleus of the cell and turning on the cytokine storm. I went back into the filing cabinet, pulled out our manuscript, attached a copy of the Harvard paper, and sent it back to the same journal, along with their rejection. I wrote nothing else. They published the article a few months later, exactly as it had been written years earlier.

With the DNA methylation project, it was a different story.

Baylin did not like to submit research which had any unanswered questions, and as a result, often held back any anomalous findings. If the author could not answer all the questions to the satisfaction of the journal, the journal would often reject the paper, and not allow it to be resubmitted at a different time.

But Frank stood firm and said we had to submit all the data, even the parts we did not fully understand.

That paper took several years of additional work to get published. We performed many experiments, using different techniques and technologies, and the results all supported our original findings. I thought the additional experiments were a tremendous waste of time.

In 1998, the paper was finally published in an excellent journal, *Molecular and Cellular Biology*. That only happened because of Frank's bulldog tenacity and brute force, and Steve Baylin's patience and creative writing. That original paper on infection of a monocyte cell line raised far more questions than it answered and led to ground-breaking discoveries in four different fields of science.

By comparison, it was simple and fast to publish my research creating the first infectious molecular clone of HTLV-1. My post-doctoral training in molecular virology was completed in less than two years, but the questions raised by my PhD thesis work took another several years to publish as we awaited, or created, the technological advances necessary to support the biological observations.

* * *

As a staff scientist back in Frank's lab in the mid-1990s, my work on inflammatory cytokines and monocytes was key to understanding an emerging, highly dangerous viral infection, Ebola.

One project I was asked to collaborate on was to determine the difference between two strains of Ebola, the "Zaire" strain and the newly categorized "Reston" strain. These studies were being conducted in the Biosafety Level 4 laboratory at the US Army Medical Research Institute of Infectious Diseases (USAMRIID), just across the baseball field from our labs at the NCI.

I conducted the studies in monocytes as a logical starting point based on my PhD thesis research. I reported the results in an abstract showing the differences in pathogenicity (disease causing) between the Zaire and Reston strains of the virus.

To put it plainly, the Zaire strain was extremely dangerous, killing by creating an inflammatory cytokine storm in primary monocytes. But the Reston strain created no cytokine storm in primary monocytes and was thus of little danger to the public.

However, the corrupt, old boys' club of science, Robert Gallo, Tony Fauci, John Coffin, and Harold Varmus, to name a few, didn't like the story emerging from our data. We were challenging the narrative, just as we'd done when we said that HIV infection didn't have to cause AIDS. Simply stated, there could be HIV infection with no AIDS, and a diagnosis of AIDS with no HIV infection. It was all about the response of the immune system.

Similarly, the Reston strain of Ebola was benign.

* * *

Frank's *frankness* got him into trouble with his superiors in 1998, although in typical government fashion, it wasn't a direct assault. Instead, it was a sneak up from behind move to stab you in the back, and hope you bled to death.

By this time, Frank was a lab chief and had several principal investigators working under him.

They initiated an investigation of Frank's lab, and although they found nothing wrong, they offered all of the principal investigators the option to leave and start their own independent laboratories. You see how sinister the power structure can be.

No, we didn't find anything wrong, they'd say to his investigators. But, if any of you want to leave, we'll help you get started in your own lab, and we'll take resources, money, and equipment from Frank's lab to help you get started.

Part of their campaign was starting a rumor that I was having an affair with Frank.

Which was humorous because, despite the fact that I had boyfriends, I was also highly competitive, both in science and in our intramural softball games, which led to another rumor that I was a lesbian. When I complained about this attempt to marginalize my accomplishments by saying I was having an affair with my boss, some of my female colleagues replied with the deadpan response, "Well, at least you're not gay."

And I hope this doesn't sound self-serving, but I enjoyed decades-long relationships with a few absolutely gorgeous, principled, and intelligent men. We simply were never in a time and space to marry. Our goals diverged but our love remains strong to this day. It's been my experience that strong men love strong women. It's the sniveling cowards, the men without a spine, who are the most misogynistic. I believe the vast majority of men do not have trouble with strong personalities, be they male or female.

This government proposal had the possibility of stripping Frank of all of his principal investigators, essentially decimating his laboratory. I'm sorry to say that all the male principal investigators took the opportunity to start their own lab.

But two of the three female principal investigators said, "No, I'm staying with Frank." Was that because he treated them with integrity and respect, and they knew they were unlikely to find that anywhere else? Frank was not simply an administrative lab chief. He was a mentor that a woman could trust.

All of this came to a head after a softball championship after which I'd probably had a few too many beers. The powers that be were also telling the principal investigators who chose to leave that they could take our laboratory equipment. There was this one technician talking about what he might take from the lab, and I drew a line in the sand. I said, "Over my dead body will one piece of equipment leave our lab." I freely confess I was emotional, and I may have also been crying over the injustice of it.

This technician, who also happened to be in a wheelchair, took this as a physical threat. It was not. The other person who overheard me, a female technician named Cary, whose husband was on the team, backed me up.

I was called into the office of Peter Fischinger, who was head of the contract lab. Frank told me once he'd told Fischinger something about Gallo that he'd told no one else. Within an hour, Gallo called Frank with the exact story. I knew not to tell Fischinger anything that could be used against me. Although the claim I'd threatened physical harm was dismissed, I was required to take a "sensitivity training" course and go see a psychologist a few times to deal with my "anger management" issues.

I'm not sure if the sessions helped me, because even today, if you try to steal my stuff, I'm still going to defend my territory with every ounce of strength I have. It's in my blood, and it's how I was raised. I knew about the "Trail of Tears," forced upon the Cherokee Nation long before they ever mentioned it in school. Take the guns away from the men, and then force the people on a long march where up to fifteen thousand are estimated to have died. No one would ever deceive me and steal my work to perpetuate fraud on the innocents, as had been done in previous centuries. Not on my watch. I plead guilty to having said more than once, "Over my dead body." I meant it then, and I mean it now.

Yeah, I know, it's a character flaw.

But I don't ever remember claiming I was perfect.

* * *

The attempt to undermine Frank's lab also came at a time of great instability at the NCI.

While Nixon had set up the structure under which a few powerful scientists could control the research community and demoted a bunch of government researchers to contractors for hire, Reagan put the system on steroids. Under his "Reduction in Force," or RIF plan, government scientists often found themselves working in the same office, in the same lab, but with less money, no benefits, and no employment rights. This is what the government does when they want to hide a bunch of money so it can go to the higher-ups, while at the same time creating an entire group of second-class citizen scientists.

The previous head of the Lab of Anti-Viral Drug Mechanisms, along with everyone else in the world-renowned contract lab, had finally had enough of this nonsense and realized they could make much more money in private industry. Most of the Lab of Anti-Viral Drug Mechanisms went to work for Celgene in sunny San Diego, California.

You really couldn't blame them.

But it left the Lab of Anti-Viral Drug Mechanisms such a ghost town that they actually closed the lab while they figured out what to do. Peter Fischinger had a big problem with everybody leaving arguably the most successful lab under his direction.

I was in Peter's office being reprimanded for supposedly threatening that technician when I piped up and said, "I can run that lab! I've worked with them for many years."

That's right. One minute I was in his office for allegedly threatening to beat up a guy in a wheelchair and the next I'm volunteering to rebuild a lab decimated by Reagan's short-sighted policies.

I got the job.

For the first few weeks I called myself the eighty-thousand-dollar washerwoman because my primary job was to clean out the lab which had been left a ghost town. Then I had the privilege of hiring the entire team of researchers. Later, many commented that our lab's personnel looked like the United Nations. I had researchers from China, Mongolia, and a Black woman from British Trinidad, and they were just great. I developed the study design to pursue mechanisms and treatments for AIDS-associated malignancies, starting with the thesis that aberrant expression of retroviruses was central to the immune system dysfunction and we needed to understand how to effectively intervene.

I'd hired this MD/PhD researcher from China, but quickly ran into a cultural roadblock with him. Even though I was head of the lab, apparently since I was a woman, he would not take direction from me.

At least when the sexism is that blatant, you can start to design a work-around.

After about four or five weeks of a failure to communicate, I came up with a plan. On Friday, I'd type up the previous week's work, what I wanted to do in the following week, then I'd go over to Frank's office for a few hours on the weekend, explain what I wanted to do, then Frank would type it up. Then Frank would sit in on our Monday morning meetings, as if Frank was the real boss, and I was just part of the team.

I'd go through what I wanted to do, who should engage in what research, I'd look at Frank, and he'd say, "Yes, that's what should be done."

Being given direction by a man did the trick. Even after he went back to Shanghai, and I left the NCI, we published a few papers together. To this day, we hold a mutual respect and admiration for each other's work.

One of the things we accomplished during this time was to make extraordinary progress against Non-Hodgkin's lymphoma. We were able to use multi-plex RNA profiling in order to identify targeted drugs and distinguish these drugs based on the patterns of expressed cytokines. Another question we were investigating was why people who had glioblastoma tumors in their brains were showing the presence of cytomegaloviruses, as well as other retroviruses in their brains. Yes, I believe that retroviruses are also involved in many brain tumors.

There was one research project which I was sorry I wasn't able to bring to completion. We had identified what we thought to be a novel herpes virus, which we tentatively named HHV-9. It seemed to be an especially nasty virus, having the ability to transform B-cells, which is exceedingly difficult to do, using Epstein-Barr Virus (EBV/HHV4). We had a strain of EBV that was highly transforming, which I now believe had XMRV env (envelope) integrated from a lab recombinant. Frank rightfully asks me where I get the data for this speculation. My answer to my mentor is there is none because after I left the lab all the samples were autoclaved. The only support I have is the JY cell line paper I presented in my 2013 talk at Dr. Derek Enlander's conference on November 13, 2013.

I'd bet money that there was also XMRV integrated into that herpes virus, and that it wasn't novel at all, but a tissue culture created recombinant. That is, there are Epstein-Barr Virus strains circulating which have retrovirus integrase, envelope sections, or reverse transcriptase, which turns them into para-retroviruses, in which they have many of the properties of a retrovirus. For example, I believe most hepatitis B infections are para retroviruses, caused by contaminated cattle (bovine) albumin, used to maintain these cell lines.

Frank is definitely right that the data isn't there to support my speculation, but I bet someday I'll be proven right.

* * *

As I was setting up the new lab, I went to a Gordon Conference in Ventura, California, March 14–20, 1999, which is where to my great, good fortune, I met my husband, David. We met the day I arrived, March 14. Frank had told me that the sunsets in Ventura were spectacular. So, before the conference started at 7 p.m. that evening I wandered out to the promenade to watch the sunset. As I looked straight out to the ocean and the Ventura pier on a beautiful clear evening, I couldn't see the sun. In my mind I was having a conversation with Frank in which I was saying, "Very funny, Frank. How can I watch the sunset if I cannot see the sun?"

As if in answer to my silent question, David walked by and said, "If you're looking for the sunset you have to go to the end of the pier. I'm going that way if you'd like to join me."

While we walked to the end of the pier, David quizzed me about the conference. I talked and talked about HIV drug development as we watched the sun set over the Santa Barbara hills to the north, a truly spectacular sight.

After the sun had set, David asked if I'd like to go out for sushi or ice cream. I explained that the Gordon conference required full participation in meals for attendees, and politely declined. It was clear that David was older than me and though I didn't flatter myself that he was asking for a date, I made it clear my passion was science.

A few days later, during an afternoon poster session, I was having a heated discussion with Dr. Peter Duesberg, a Berkeley professor, about whether the HIV virus caused AIDS. "Peter," I said imploringly. "I know HIV doesn't cause AIDS. [Though we all knew it was PART of the pathogenesis] But my job is to develop therapies for AIDS-associated cancers. Clearly, when the CD4 T-cells can't do their job, certain types of cancer will inevitably develop."

Duesberg would not back down. I was cornered and looking for an escape. Just then I saw David walking down the promenade, where we'd met a few days earlier.

I said, "Pardon me, Peter, an old friend and I are supposed to go for a little walk before the evening session."

I practically sprinted away from Peter and called David's name.

When I reached him, I took his arm, then leaned in and said, "Don't turn around. Just keep walking. You said there's good sushi around here? I have to be back by the 7 p.m. session. It's the most important of the conference."

David didn't miss a beat.

Several hours later, after lots of sushi, Sapporo, and sake, David had heard all about AIDS drug development and misogyny at the NCI.

At the conference, I'd also met some great researchers from the University of California, Los Angeles, and we set up some collaborations to study herpes viruses, AIDS-associated malignancies, and mouse gamma viruses. In addition to being some great research, it also allowed me to spend more time with David.

David proposed to me on April 1, 2000, my birthday, and yes, I know, it's April Fool's Day. I always say God has a sense of humor.

The irony is that Ventura was arguably Ground Zero for what I believe has been the greatest crime against humanity, the COVID-19 plandemic. While Frank argues the Holocaust and Native American genocide were greater atrocities, my belief is that COVID-19 was a planned mass murder to cover-up the tens of millions of people infected with animal retroviruses because of a contaminated blood supply and contaminated vaccines. Leading this genocide was Tony Fauci, who for decades has consistently ignored the evidence of human retroviruses other than HIV contributing to AIDS.

Frank believes this statement is far too strong and the existing evidence is far too weak to say this with certainty. I stand by these statements because not a single assertion we've made in our books has been shown to be wrong. I believe no statement in this book will be shown to be wrong, either. I appreciate that Frank has family, whom he believes may find their careers adversely affected by such speculations. I believe it is imperative for all of us to step out with boldness to protect every human life at risk in this age of corruption.

In the summer of 2011, as our investigation of XMRV was making me aware of many threads in multiple diseases, I came to believe that those who died in the first wave of the AIDS epidemic were those infected with both HIV and XMRV. XMRV crippled key players in the innate immune response, the plasmacytoid dendritic cells, which make up more than 95 percent of the body's type 1 interferons and the natural killer cells, which kill cancer cells, as well as virally infected cells. Even now, it's difficult for me to consider how many lives could have been saved if scientists like Peter Duesberg and their legitimate questions had not been censored.

The destruction of the scientific careers of those who question the prevailing orthodoxy remains an ongoing threat, not just to those brave scientists, but to all of humanity.

* * *

David and I were married on October 7, 2000. I then spent several months commuting between David in California and the Lab of Anti-Viral Drug Mechanisms at Fort Detrick in Frederick, Maryland. While in California with David, I interviewed with a company called EpiGenX, which impressed me, and they offered me a job as their head of cancer research.

My last day working as the head of the Lab of Anti-Viral Drug Mechanisms was May 11, 2001, and on May 14, 2001, I started work at EpiGenX. I had been the head of the lab for more than two years and I felt I'd done a great job rebuilding it and placed it on a solid footing. This is the recommendation letter which was sent on my behalf to EpiGenX by one of my long-time colleagues, Stephen B. Baylin:

Dear Ms. King:

It is my great good pleasure to recommend Dr. Judy Mikovits for a position at EpiGenX. Simply put, she is a first-rate scientist who works as hard as anyone I have known. I had the good fortune to collaborate with her on the work she has done and published in DNA methylation, and it was a great experience. She conceived and tackled largely with her own hands a difficult and novel problem regarding effects of the HIV virus on the epigenetic control of interferon expression. A superb MCB paper culminated this work and I think contains data that will prove unexpectedly critical for understanding the clinical aspects of HIV and possibly some therapeutic and prognostic issues as well.

Judy loves science, is creative, and tenacious when pursuing an exciting finding or direction. She has a fine career both in evolution and ahead of her. You would do well to try and recruit her.

Stephen B. Baylin
Associate Director for Research
Director, Cancer Biology Division
Ludwig Professor of Oncology
The John Hopkins Comprehensive Cancer Center[2]

There you have two recommendation letters, one which summed up about twenty years of my life from being a lab technician, through graduate school, and my post-doc career. The second indicated that after approximately two years and four months of being head of a lab, I impressed people.

Later I learned that Peter Fischinger, who'd given me my shot to be head of the Lab of Anti-Viral Drug Mechanisms, apparently believed the rumors that I was sleeping with Frank. Of course, there couldn't be a brilliant man and a woman with some brains of her own working together and not have a little something going on the side.

By the way, Frank does not believe this scenario. He believes they separated us in the belief we would fail.

I do not see the two beliefs as incompatible.

Once again, Frank and I are probably both right.

CHAPTER TWO

Interferon as an Anti-Tumor and Antiviral Agent

The cover of the March 31, 1980 issue of *TIME* magazine changed my life. The screaming headline was "INTERFERON — The IF Drug for Cancer," along with an extreme close-up of a syringe with a single reddish-pink drop hanging from the tip.

The article gave the history of interferon up to that point. But my concerns were much more down to Earth. I was a senior at the University of Virginia when that story came out. Even then I vividly remembered my favorite grandfather, the one who'd taught me to love baseball, had died from lung cancer when I was twelve years old. I had vowed then as a young girl to take on cancer as my life's mission.

Another great influence in my life was the University of Virginia, founded by Thomas Jefferson, and its totally student-run honor code. A student could not lie, cheat, or steal. If such an infraction was committed, the sole remedy was permanent dismissal from the University. To some that may sound harsh. But to me it seemed exactly how the world should function. In its explanation of the code, the University stressed how such a strict honor code created communities of trust. To see an honor offense committed and not report it was an honor offense. It was and still is the only one of its kind in the United States. It's just as our parents taught us.

Our word was our bond.

We honored our word and each other.

It was a wonderful community of trust.

Maybe that has blinded me to much of the corruption and lies in the world around me. But when I read that *TIME* article in 1980, I read it with the belief it was the leading edge of medical science. More than forty years later, I still believe in the essential accuracy of what I read. It is maddening to me that we have not made more progress. This is how the article vividly described the development of cancer.

> It can start in just one of the body's billions of cells, triggered by a stray bit of radiation, a trace of toxic chemical, perhaps a virus or a random error in the transcription of the cell's genetic message. It can lie dormant for decades before striking, or it can suddenly attack. Once on the move, it divides to form other abnormal cells, outlaws that violated normal genetic restraints. The body's immune cells, normally alert to the presence of alien cells, fails to respond properly; its usually formidable defense units refrain from moving in and destroying the intruders.[1]

Don't you feel that much of the last forty years of environmental protection can be summed up in that single paragraph? Who doesn't worry about human health, either for themselves or for the greater good of humanity? I believe every person reading this book wants humanity to experience the maximum amount of good health and for people to lead long and productive lives. We worry about things like radiation (from our cell phones), toxic chemicals in our food, air, and water, viruses, and whether our genes are functioning for our optimal health.

I thought the section from *TIME* was similarly excellent in describing the process of cancer development. For some reason, the aberrant cancer cell is created, then, in some process we do not fully understand, tricks the immune system into not responding. I believe every living thing has intelligence. Even though many question whether viruses can even be classified as living things, I am convinced there is an intelligence to their actions as well.

They hijack the cellular machinery to reproduce and spread throughout the body and yet we do not understand how they get the immune system to stand down. Does interferon act as a sharp-eyed policeman, observing the attempted assault, and calling in reinforcements? Or does interferon itself go charging in for the assault?

We now know from studying our endogenous virome (viruses which have become integrated into our genetic code) that the answer is that both are true.

The article did a good job of summarizing the history of interferon, how it was first identified in 1957, and then how a Finnish virologist, Kari Cantrell, was able to start synthesizing larger amounts. The 1980 *TIME* article continued:

> Still, researchers now had enough interferon to move studies out of the laboratory and into the clinic. In 1972, virologist Thomas Merigan of Stanford University, and a group of British researchers began studying IF's effect on the common cold. Soviet doctors were claiming success in warding off respiratory infections with weak sprays of IF made in a Moscow laboratory. Merigan and his colleagues gave 16 volunteers a nasal spray of interferon one day before and three days after they were exposed to common cold viruses. Another 16 volunteers were subjected to the same viruses without any protection. The results seemed miraculous. None of the 16 sprayed subjects developed cold symptoms, but 13 of the unsprayed did. There was one catch: at the IF strengths that Merigan used, each spray cost $700.[2]

The problem was the cost of the synthesis. Bring this cost down with new technologies and as the article detailed, the possibilities were endless for every person to prevent infectious disease, and maybe even cancer. The article continued:

> In the years since, Merigan and his Stanford team have successfully used IF to treat shingles and chicken pox in cancer patients. In other studies, IF has prevented the recurrence of cytomegalovirus (CMV), a chronic viral disease that sometimes endangers newborn babies and kidney transplant patients. Israeli doctors have also used IF eye drops to combat a contagious and incapacitating viral eye infection known as "pink eye." Researchers are now trying a combination of IF and the antiviral drug ara-A in patients with chronic hepatitis B infections. Interferon investigators have high hopes that the drug will be equally active against other viral diseases.[3]

The concept that interferon might also be effective against cancer may have occurred spontaneously to several researchers as the work of Isaacs and Lindemann was confirmed. After all, it had already been shown that some animal cancers were caused by the polyoma virus. Though no human cancer virus had yet been identified, some tumors seemed linked to viral infections.[4]

This made an enormous amount of sense to me. While an acute viral infection will put a short-term stress on the body, chronic viral infection has the capability to exhaust and degrade the immune system so that cancer cells can replicate out of control.

Can you understand why I was so excited when I read this *TIME* magazine article? Interferon was showing promise as a treatment for the common cold, shingles, chicken pox, pink eye, and was being investigated for its effectiveness against chronic hepatitis B infection.

Just after graduation, I took a break from the job search for the Memorial Day holiday weekend when I saw an ad in the *Washington Post* for a technician to purify interferon for the National Cancer Institute. I couldn't believe my eyes and applied immediately. It was exactly the job I'd hoped for since I did not get into medical school.

I thought I could take a year off, cure cancer, and then I would atone for those Ds in Organic Chemistry, as I could help get the cost of manufacturing interferon down low enough so that everyone could benefit.

I applied, got called for an interview, and began my dream job on June 10, 1980.

You might say interferon was my first great love in science.

Most of that first year at the NCI I was working on several different projects. I worked on the fermentation chemistry team that purified interferon in large amounts from large scale fermentation cultures of the Burkitts' lymphoma cell line, Namalwa. I also purified other natural products from plants, like the breast cancer therapy, Adriamycin, called the red devil or toyocamycin, and sangivamycin nucleoside analogues which were active against human cytomegalovirus.

Later, these interferons and natural products were used clinically for the treatment of cancer and HIV/AIDS.

* * *

After the 2020 publication of our book *Plague of Corruption*, I was contacted by a long-time veterinary professor, Dr. Joseph Cummins, who spent much of his career studying interferon in animals and humans.

I learned Dr. Cummins was a legendary figure in veterinary medicine, the holder of sixteen US patents and author of sixty peer-reviewed scientific articles, as well as a vital contributor in the field of HIV/AIDS research with leading scientists. Cummins did his undergraduate work at Ohio State University, served as a captain in the US Army, and received his doctorate in microbiology from the University of Missouri. In 1999, Dr. Cummins was named a Distinguished Alumnus from Ohio State University. He was in the very same department at Ohio State where my graduate school friend Kathy went as a professor to start her career after her postdoctoral studies in the lab of Nobel Laureate Howard Temin, discoverer of reverse transcriptase.

Cummins had put together a book about his decades of work in interferon, but thought it needed some work, and did I know a good writer? I put him in touch with my coauthor, Kent Heckenlively, who spent several months getting the book into shape, and it was published in January 2021 under the title *The Case for Interferon: How a 1980s Cancer Drug Might Be the Wonder Therapy for the Twenty-First Century.*

Maybe there are a few truly independent entities which will honestly look at what original thinkers have to say. I heartily encourage you to pick up a copy of this book.

Unfortunately, my experience of interferon treatment of retrovirus associated diseases, HIV/AIDS and XMRV/MECFS, suffered from much of the same institutional corruption that Joe experienced.

* * *

I know the public likes to think of scientists as living completely in our heads, but the best of us are also engineers at heart. We like to come up with new approaches to doing things or understanding how new technology can help us answer vexing questions.

Everybody else was looking for interferon in the blood.

But when you understand the mouth and nostrils are the gateway to the interior of our body, you realize that's exactly where Mother Nature would place some of her best guardians. The battle is being fought first in the respiratory system, and then if that fails, will often move to the bloodstream. Purifying interferons from the continuously growing Namalwa cell line was a significant advance in technology and could and did make the low doses needed for therapeutics affordable.

Several articles stated that while cancer fighting has been grabbing all the headlines, interferon's potential as a protection against viruses remained powerful. The protein blocks the action of every virus it has been tested against in the lab. It seems to be able to stop some common cold viruses dead in their tracks, and to offer hope for eliminating the dangerous carrier of chronic viral hepatitis.

However, we are constantly told more testing must be done in these areas before interferon's antiviral potency can be fully judged.

Remember my central belief: that all of nature, plants, and animals are given by God and, as it says in Genesis 1:31, for our good health.

So, why did this effort, started with great fanfare in the 1980s, not come to fruition?

The problem was not with God.

I believe the problem was human intervention.

* * *

In Joe's book, there was one story that absolutely mesmerized and haunted me, making me think back on the treatments for XMRV/MECFS, which were denied.

There was Jonathon Kerr out of the United Kingdom, who denied the use of type one interferon for nasal usage in ME/CFS patients.

The biologic therapy, Ampligen, a mismatched double-stranded RNA molecule with immunotherapy and antiviral properties, has been denied approval by the FDA for decades.

There was Robin Weiss, who along with Coffin collaborator Jonathon Stoye, denied the UK trial of type I interferon in ME/CFS patients, while Simon Wessly was pursuing behavioral modification.

Yet in 2010, twenty-five years after it was shown that type I interferon could silence retroviruses, Frank presented the data showing XMRV's blocked production of type I interferon from plasmacytoid dendric cells, suggesting low dose interferon could stop the transmission of retroviruses. To this day it is still being denied for millions of patients with XMRV/MECFS and their families who may be at risk.

In the spring of 1986, Joe got a call from another veterinarian, Dr. Buddy Brandt, asking how he might dose a large dog with a tumor, using Pet Interferon Alpha. Joe thought the weight reported was at the upper limit for a dog, but still within a believable range.

In truth, there was no big dog.

The "dog" was the veterinarian himself.

Dr. Buddy Brandt had undergone surgery in November 1983 and required multiple blood transfusions. In February 1986, he was diagnosed with AIDS-related complex. Although Joe did not realize Buddy was treating himself with the Pet Interferon Alpha, when he started to get good results, he told Joe. Eventually, Joe and another scientist, Val Hutchison, submitted what was essentially a case report on Buddy to *The Lancet*, and it was published in 1987:

> In late 1983, a 40-year-old man underwent open-heart surgery and complications necessitated many blood transfusions. In early 1984 he was admitted to hospital with pneumonia, and this may have been his first AIDS-related illness. Subsequently, herpes, genital warts, diarrhea, herpes simplex, mouth ulcers, respiratory infections, and weight loss became chronic, and in February 1986, AIDS-related complex (ARC) was diagnosed.
>
> Our patient was a veterinary surgeon. Success with oral human interferon-alpha (IFN-*a*; pet IFN-*a*, Amarillo Cell Culture Co) in feline leukemia and his chronic weight loss and depressed T4 lymphocyte counts prompted him to experiment with IFN-*a*. Self-treatment was followed by a rise in T4 lymphocyte count from 153 to 319 /ul, regression of genital warts, improved appetite, and weight gain.
>
> The interferon treatment was discussed with his physician who discouraged its use, and the treatment was discontinued; his T4 lymphocyte count decreased within six weeks to 146 /ul. His physician prescribed ribavirin for the next 8 months, and during ribavirin treatment the patient experienced weight loss and anorexia, and his T4 count varied from 146 to 218 /ul. In January 1987, his condition deteriorated from ARC to AIDS, according to his primary physician.
>
> At the end of January, 1987, he began a course of low dose (2-4) units/ kg body weight per day) oral human IFN-*a*, and over the past 11 months his appetite has returned, he has gained weight, his herpes simplex and mouth ulcers have disappeared, the genital warts have regressed and recurred, and T4 counts have risen from 210 to 520 /ul. He feels better than he has for 3 years and has been able to maintain the quality of his life and has not been off work for those 11 months . . ."[5]

That letter in *The Lancet* brought worldwide attention to Joseph Cummins and his low-dose interferon treatment. During this time, Joe had also been able to create a partnership with a Japanese company, Hayashibara Biological Laboratories, to produce interferon using Joe's techniques, which they found to be of exceptionally high quality.

In early 1989, Joe had become so well-known that he was asked to fly to Kenya and meet with Dr. Davey Koech, who was director of the Kenya Medical Research Institute (KEMRI) to discuss plans to provide interferon for a trial of HIV/AIDS patients. This research was completed, and the dramatic results were published in the journal, *Molecular Biotherapy*, in June 1990. Here's a section from that article on the results in forty-two patients:

> The CD4+ cell count increased in all but 4 of the treated patients; the average increase was 291.1 cells/cu.mm3 after two weeks of natural human interferon treatment. Concurrent with increases in the CD4+ cell count were increases in the KPS [Karnofsky Performance Status – A measure of a person's ability to perform the daily activities of life. The range is from 1 to 100, with 100 being normal.] Patients experienced dramatic symptomatic relief and all of the patients except 3 had a KPS of 100 after 2, 4, or 6 weeks of human interferon treatment.
>
> The average weight gain of patients after six weeks of oral human interferon therapy was 4.6 kg [a little more than ten pounds] and is in contrast to loss of appetite and other toxic effects associated with high dosage parenteral usages of human interferon. The low dosage and oral administration of human interferon differ in comparison to other human interferon treatment regimens that have been tested in ADS and HIV-1 seropositive patients.[6]

At the time the study was published, the CD4+ cell count was generally thought to be the most accurate indicator of a person's health, so the increase in a positive direction was an unmistakably positive sign. I'd also suggest that the increase in being able to attend to the daily activities of life was of great significance. The AIDS drugs which were being given at the time, while having some positive effects, were often causing appetite suppression and dangerous weight loss. Interferon seemed to be an all-around winner.

But the most surprising finding was the apparent disappearance of HIV in a few of the patients, or at least beyond our ability to detect it at that time.

> Four patients turned negative for both ELISA and Western Blot by the second week after treatment, and another 4 patients turned negative by the fourth week. These individuals have since remained negative during subsequent follow-ups and have continued to be asymptomatic. This is the first observation where there has been "seroconversion" on both ELISA and Western Blot and clinical improvement resulting, apparently, from therapy.[7]

But the 1990 study from Kenya was not the first to show the positive benefit of interferon, especially when used in combination with other AIDS drugs.

This is how the *New York Times* described an earlier study when interferon was used in combination with AZT, an early AIDS drug:

> In 1990, AZT was the only drug approved for the treatment of acquired immune deficiency syndrome, but as many of half the patients cannot tolerate the recommended doses because the drug damages their bone marrow.
>
> In the study, 22 men infected with the AIDS virus took low doses of the two drugs for 12 weeks and the drugs apparently worked together to inhibit the viral infection. For example, six men had viral proteins in their blood when the study began, but those proteins disappeared in three of them. Six of the 22 men no longer had viruses that could be isolated in their blood.[8]

Because of my PhD thesis, which was in progress, I appreciated the importance of these findings. If we keep the virus latent or silent, the patient lives just like the rest of us. Yet, not only was this safe, low dose treatment denied general use for prevention and treatment of HIV/AIDS, but it has been denied in other diseases associated with retroviruses including ME/CFS and ALS.

The article went on to quote a name which has now become well known to virtually all Americans:

> Dr. Fauci said the growing evidence that interferon is effective against AIDS patients who could not take AZT because of its side effects should consider taking alpha interferon instead of, or in combination with AZT.
>
> "I believe that alpha interferon is very promising," he said. "Independent studies in the United States and Canada have clearly shown it is effective against Kaposi's sarcoma and that it had a significant effect," against the AIDS virus.[9]

Now, there are some caveats to be certain. Joe always said that interferon should be used in low doses. Sometimes even the low doses that Joe suggested also needed to be cut in half, as was needed in the Kenyan study under Dr. Koech. It may have been that their immune system was in such a weakened state that it simply needed to be brought along a lot slower.

But then everything started to collapse, not because of the failure of the therapy, but because of human institutions.

But interferon keeps finding its way back into the news.

In April 2020, a team from the University of Texas reported in the *Journal of Antiviral Resistance* that interferon was effective against SARS-CoV-2, the virus that causes COVID-19. The researchers found:

> Our data clearly demonstrated that SARS-CoV-2 is highly sensitive to both IFN-*a* and IFN-*b* in cultured cells, which is comparable to the IFN sensitive VSV. Our discovery reveals a weakness of the new coronavirus, which may be informative to antiviral development . . . Our data may provide an explanation, at least in part, to the observation that approximately 80% of patients actually develop mild symptoms and recover. It is possible that many of them are able to mount IFN-a/b-mediated innate immune response upon SARS-CoV-2 infection, which helps to limit virus infection/dissemination at an early stage of the disease.[10]

Let's discuss the importance of these findings. SARS-CoV-2 is very sensitive to interferon, both type A and type B. This may explain why 80 percent of those who contract the virus develop mild symptoms and recover. Their immune system is properly producing interferon in the amounts needed to successfully fight off the virus. I understand it's one thing to show these results in culture, but what about in human beings?

We saw the dramatic results in HIV/AIDS, especially when interferon was paired with another antiviral. Well, researchers in Hong Kong did that very study and published their findings in *The Lancet* on May 30, 2020:

> In this multi-center, randomized open-label phase 2 trial in patients with COVID-19, we showed that a triple combination of an injectable interferon (interferon beta 1b), oral protease inhibitor (lopinavir-ritonavir), and an oral nucleoside analogue (ribavirin), when given within 7 days of symptom onset, is effective in suppressing the shedding of SARS-CoV-2, not just in nasopharyngeal swabs, but **in all clinical specimens** [bold and underline added], compared with lopinavir-ritonavir alone.
>
> Furthermore, the significant reductions in duration of RT-PCR positivity and viral load were associated with clinical improvement as shown by the significant reduction in NEWS2 and duration of hospital stay. Most patients treated with the triple combination were RT-PCR negative in all specimens by day 8. The side effects were generally mild and self-limiting.[11]

Makes perfect sense, since the SARS-CoV-2 spike protein contains HIV Env (envelope), XMRV Env, and SARS receptor binding domains. Syncytin

is the name of the envelope protein exp
retrovirus, HERV W (the 7C10 antib
with all known polytropic and xenotr
HIV Gp120 sequences were detecte
firmed by our colleague and No
presented the data supporting th
and my statements have never b

Yet the denial of effective
XMRV-infected communities, while Fauci
"long-haul COVID" and forces the 12 percent can
inoculated, rather than simply providing low-dose, type I interfer
or ivermectin, is truly a crime against humanity.

Furthermore, the significant reduction in duration of the RT-PCR positivity was associated with clinical improvement as shown by the significant reduction in symptoms and duration of hospital stay. Most patients were negative by the eighth day of treatment and side effects were mild.

A triple combination of drugs, with interferon at its core, was enough to make most patients RT-PCR negative in all specimens within eight days. This is a tremendous breakthrough. Now that the fear has passed, let's look at the situation objectively. Eighty percent of those exposed to SARS-CoV-12 will develop only mild symptoms of COVID-19. The 20 percent who do develop symptoms now have a treatment protocol.

In many instances, an obstacle can present an opportunity. Since the regulating agencies have concluded that interferon at a low dose has no effect, it can legally be sold as a food supplement. Food supplements can be used to support "general health" and no other claims are made about this product. Got it? I've shown you the published research, but am making no claims other than it may help with your general health. That should satisfy the lawyers.

More than forty years after I first fell in love with the promise of interferon, that torch still burns bright. The world owes Joseph Cummins a debt of gratitude.

CHAPTER THREE

The Gatekeepers in Science

I'm often asked, "Judy, how do you put up with what they've done to you for the past ten years? Has this completely disillusioned you about science?"

My answer to them is, "I've been fighting with this misogyny from the beginning of my time in science more than forty years ago." Sometimes the players change, but often the pattern is just the same.

I've mentioned before in my books, as well as earlier in this one, that I was first hired by the NCI as a technician on June 10, 1980 to work in the fermentation chemistries program at Fort Detrick in Maryland. In effect, we were the hired guns, the technicians in the lab, figuring out how to complete the studies of the god-like "principal investigators." In the public health hierarchy, you should think of the principal investigators as akin to the general of an army. As lab technicians, we were like the boots on the ground. The general might say "I want you to take that hill! Get ready to charge!" and we'd be expected to do it.

But if you're a smart soldier, you take a look at the hill, note the enemy fortifications and the location of machine gun nests, and say to yourself, "If we charge up that hill, we're going to get mowed down by those machine guns."

You then go to your sergeant and say, "Sir, we're going to get slaughtered if we charge up that hill. But there's a steep canyon on the other side that's unguarded and I think we can surprise them."

Hopefully your sergeant has an ounce of brains and thinks, *Do I want to be praised for executing the order exactly as the general gave it, fail to take the hill, and get all my men killed? Or do I want to take the risk the general gets mad I didn't fulfill the order as he gave it, but I capture the hill and save the lives of my men?*

The first project I worked on was purifying interferon, which at the time was being investigated as a highly potent anti-cancer and anti-viral therapy.

Late in 1982, our team was assigned a project which would put me in direct contact with many of the most important players in public health. Dr. Robert Gallo wanted us to provide his lab with thirty grams of human T-cell leukemia virus (HTLV-1) from a two-hundred-liter (about fifty-two gallons) culture of Hut-102 cells.

Hut-102 cells were drawn from the peripheral blood of people with adult T-cell leukemia. HTLV-1, the first identified disease-causing human retrovirus, was discovered in 1980 (HIV would be discovered in 1983) and associated with the deadly leukemia, found most commonly in Japan, of adult T-cell leukemia.

To extract thirty grams of infectious HTLV-1 virus from two hundred liters of cell-culture, we required an open-air centrifuge known as the AK centrifuge.

My boss Mark and I immediately realized that nothing was known about the potential for this virus to become airborne and infect the technicians, particularly a few who were pregnant. I understand I've had a PhD after my name for more than thirty years, but in my heart, I'm a technician, just like Frank. I'm never happier than when I'm working in a lab. Little was known about the transmission, infectivity, or mechanisms by which this virus caused disease. Mark and I wrote a letter to our superiors requesting additional time and resources to complete the assignment with proper safeguards.

Later, I'd learn, it was Robert Gallo—who fired Frank, the discoverer of HTLV-1 and IL-2, because he was "getting too much credit"—who himself made the decision to use Reagan's "Reduction on Force" mechanism to fire the members of our team for questioning the safety practices at the NCI.

* * *

Prior to his departure from the NCI under an ethical cloud in 1992 for possible criminal actions in HIV/AIDS research, Robert Gallo had more control over public health than any scientist in history, prior to the quarter-century-long march of power given to Anthony Fauci with each plandemic.

Starting with HIV and marching through SARS, bird flu, swine flu, XMRV, Ebola, Zika, and culminating in SARS-CoV-2, the supposed monkey virus never isolated from a person with COVID-19, these criminals have destroyed the economies of nations. COVID-19 was never that monkey virus, but the criminal Fauci-led plandemic, one of the most-deadly worldwide plagues in modern history.

I'd like the reader to understand that there have been just a few key scientists over the past forty years who've dominated public health. It's not a long list.

Robert Gallo, Anthony Fauci, John Coffin, Ian Lipkin, and Harold Varmus.

I speak against these men, but I also speak against a system which has allowed them to stay in positions of power for so long.

While most were enjoying the Christmas holidays in 1992 and getting ready to celebrate the New Year, the federal government quietly gave their answer to the long running allegations against Robert Gallo, hoping few would pay attention as the country was making its transition from President George Bush to Bill Clinton.

On December 30, 1992, the Federal Office of Research Integrity released a report, based on three years of investigation, which found that Robert Gallo had committed "scientific misconduct."[1] This is how it was reported in the pages of the *New York Times* on December 31, 1992:

> Dr. Gallo has faced questions about his scientific claims ever since the paper was published in *Science* magazine in April 1984. Most of his critics argued that Dr. Gallo had tried to take credit for work that French scientists were studying and claimed it as his own. At the time, the virus was difficult to isolate and grow in sufficient quantity for research . . .
>
> Dr. Gallo has denied any wrongdoing in the most vehement terms. He has also alleged that there is a conspiracy to discredit him and has asked why it is only his laboratory being investigated, and not that of Dr. Luc Montagnier, the French laboratory leader who has largely escaped detailed scrutiny.[2]

Okay, who's the conspiracy theorist in this story?

Let's put a timeline on this. Dr. Gallo takes credit for isolating the HIV retrovirus in April 1984. On December 30, 1992, eight and a half years later, the government released a report just before the New Year, when practically nobody would read it, that Gallo was a thief.

In an Islamic country, you might get your hand cut off for theft.

In the United States, you'll likely get a stiff prison sentence.

But if you're a nationally recognized American scientist, whose lies "impeded potential AIDS research progress," likely contributing to the unnecessary death of millions, you'll be able to set yourself up at your own "Institute of Human Virology" at the University of Maryland, which will list you as the founder and director and note that you're "best known for [your] co-discovery of HIV."[3]

Besides being a virus thief, the government report cited Gallo for using his role as a referee on another article in a French journal to rewrite several sentences to favor his hypothesis, four instances of scientific misconduct in his 1984 paper that he tried to blame on his co-author, poor record-keeping in his laboratory, and refusing to share his cells with other scientists to duplicate his work.[4]

And the conflict with French researchers Luc Montagnier and Françoise Barré-Sinoussi became an international incident, requiring delicate international diplomacy.

> The dispute over Dr. Gallo's claims became so linked to national scientific prestige that the Presidents of France and the United States attempted to end the conflict in 1987 when they agreed to a 50-50 split of credit and patent royalties from work with the AIDS virus and the blood test to detect it.
>
> But the issue did not go away, and Federal investigations were begun in 1989, after a reporter, John Crewdson, of *The Chicago Tribune* wrote a 50,000-word article laying out many of the charges against Gallo and his laboratory.[5]

It's one thing to get into a fight with a sibling so bad that your mom or dad have to step in. It's quite another to screw up a scientific inquiry into a disease killing millions that's so bad the leaders of your two countries have to step in to resolve the issue.

The work of *Chicago Tribune* reporter, John Crewdson, was so critical to the case that the paper's public editor, Douglas Kneeland, felt compelled to publicly comment on it.

> What has been at issue all along here is something much greater than the sum of those parts: the failure of the American government, including the Bush administration as a whole, the Departments of Justice and of Health and Human Services, in particular, and of the national scientific community to deal appropriately with the outrageous behavior of so many involved in the affair.

That behavior ranged from the repeated lies of Gallo over many years and in many respected journals in defense of his unconscionable claim to the discovery of the AIDS virus to the obvious efforts by some officials at the National Institutes of Health to cover up politically and scientifically embarrassing findings in the case.[6]

That's the way journalism used to work in America. It might have taken the *Chicago Tribune* three years to finally get the goods on Gallo, then three years for the government to do its investigation, but some measure of truth was finally told.

Compare that to *Meet the Press* and Chuck Todd taking less than a month to slander our book *Plague of Corruption* and reputation without even giving us a chance to respond.

* * *

I had some time left at the NCI until the reduction in force kicked in and I was out of a job, so I thought I'd look for other work.

As Frank mentioned in Part One, one of the great things about the NCI was they'd often have these lunchtime talks given by researchers that anybody could attend.

There was a talk being given by Dr. Joost Oppenheim, nearing fifty at the time, which seemed ancient to my perspective at the age of twenty-three. His talk was about interleukin 1, a cellular communication protein, which like interferon, was being looked at as a possible cancer treatment. I was terrified to approach him and waited until everyone else had gone, for fear my question about using IL-1 and interferon combined in a single protocol to cure cancer was a stupid one.

To his credit Oppenheim didn't laugh, but rather invited me to have a lengthier discussion about the topic over lunch in his office. That was where I'd always wanted to be, on the leading edge of medicine, developing the cures of the future.

During the discussion with Dr. Oppenheim I let him know about the reduction in force and that I was looking for a job. I was hoping he might have a position available in his lab, but he said he knew a colleague who could use a good technician. He told me his colleague was Dr. Frank Ruscetti, who was interested in studying HTLV-1 and T-cell growth factor (later designated as interleukin 2).

Years later, I learned of Dr. Oppenheim's history. Dr. Joost Oppenheim, born in 1934, is still working at the NCI as I write this in June 2021. He'd been born in the Netherlands, and as a young Jewish boy had been forced to wear a Yellow Star of David by the Nazis. For the duration of the war, he was hidden by a non-Jewish family when his parents were sent away to Auschwitz. His father was murdered in Auschwitz, but his mother survived the camp. When he was reunited with his mother after the war, they immigrated to America and he pursued a career in science.

When I learned of his history, it answered a lot of questions for me.

Rarely have I met a kinder individual.

Jo would always invite visiting fellows to his home for holiday dinners. And he did not hesitate to invite this twenty-four-year-old to chat about his research in the spring of 1983. It never occurred to me that Dr. Oppenheim had a very busy lab and might have many important things to do.

That's the difference between the real scientists and the gatekeeper misogynists of science.

For Robert Gallo, Anthony Fauci, Ian Lipkin, and Harold Varmus, it was all about what science could do for them.

For Frank Ruscetti, Joost J. Oppenheim, and Luc Montagnier, it was about the pursuit of knowledge and the ability to alleviate suffering with their discoveries.

I didn't realize it at the time, but we were both victims of Robert Gallo's megalomania. This professional jealousy was hard for me to fathom, especially as a young idealistic woman who was simply interested in advancing human knowledge and helping people live longer, healthier lives. Yes, I do ask a lot of questions, and if you give me a stupid answer, I don't care about your position in the public health hierarchy, or your distinguished publications, because it's still a stupid answer, and I won't be quiet about it. It's a scholar's obligation not only to produce knowledge, but to communicate it honestly. I don't tolerate BS or fools lightly.

As illustrated by our conversation during my job interview, detailed in Part One, I did not need to measure my words or fear asking stupid questions. No matter what he thought, Frank Ruscetti would always answer in a way that showed he welcomed the question. But make no mistake, his answer would always be an intelligent response, supported by data. Even if I was brimming with the confidence of a twenty-five-year-old and totally wrong, Frank Ruscetti would try to get me out of the embarrassing situation, even if it appeared otherwise to me.

The first months of our, as of June 6, 2021, thirty-eight-year collaboration, were very difficult for both of us. Frank's intuitive genius was clear to me from the start, just as it was likely clear to Robert Gallo. Without Frank, the discovery of HTLV-1, IL-2, IL-5,IL-15 for immunotherapy, or TGF-beta as a bifunctional regulator of the hematopoietic stem cell would have been greatly delayed.

Many a day in that summer of 1983, I'd sit in front of the biological safety cabinet well after 5 p.m. and cry because I couldn't figure out how to make a bottle of media for cell culture. I was the only kid in our family to go to college. My parents would brag about their daughter who worked at the NCI and was going to be a doctor. I could never face them or quit. I was going to have to start over and hope Frank was more than just brilliant. I prayed he was a man of honor.

I was a chemist and couldn't find a cell under a microscope. I rarely understood a word Frank said. Not just because of his thick Boston accent, but because he spoke a different scientific language. For example, I had never heard the expression "colony stimulating factor" when Frank used it in one of our early meetings.

I only knew of the original colonies in Virginia and nudist colonies. That was a question even I knew was likely too stupid to ask, and I suffered in silence. Imagine my terror when Frank casually mentioned he wanted me to "split the cells" or "feed the cells." I fed snails while a student at the University of Virginia. Some chalk, a piece of lettuce, and a square of toilet paper in a Petri dish. How difficult could feeding cells be? All of this would probably have gone much better if I could have found the cells under a microscope.

I often pretended to know exactly what Frank was saying, nodding in affirmation. In addition to his thick Boston accent and new scientific language, was his atrocious handwriting. How would I ever get this job done?

I came up with what I thought was a perfect solution. I bought myself a journal and simply wrote down every word he said. Then, I compared it to his written instructions in the hope I could translate it overnight. That allowed me to come to work the next day with an experimental protocol ready to go.

My plan worked.

I was coming to work each day with growing confidence I could make Frank happy and complete the experiments he'd assigned to me. My mom was a great secretary and she'd taught me aspects of shorthand so I could write quickly and accurately. It was a habit I've continued in my now thirty-eight-year collaboration with Frank.

Frank was so wounded by what happened to his career and reputation *because* he made two discoveries that changed all of immunology. Who knew in 2011, my dozens of personal journals would be called "intellectual property theft" and my career would end like Frank's had almost ended thirty years earlier? Now the gatekeeper scientists, whose views of the scientific world were sacrosanct, simply eliminated me and my work because it did not fit with their plans.

How do they expect me to defend my work?

It's as if the Catholic Church marched into Darwin's study, took his notebooks, and then challenged him to prove his theory of natural selection.

Frank Ruscetti should have had more postdocs, technicians and scientists working under his direction than just me. I was wounded by the Reduction in Forces canceling of my job, but more damaged by the dawning realization that science and curing cancer were not the true goals of the gatekeepers at the NCI. The primary goal of these gatekeepers was simply more money and fame.

Fortunately, Frank and I both learned to relieve the frustrations with science by throwing ourselves into competitive sports. I loved baseball and squash the most because you could win at both if you outthought your opponents. Close behind were rowing crew and ice hockey.

My career in colony stimulating factors and hematopoiesis turned out to be short-lived.

One day I came back to the lab after a lengthy lunchtime squash match to find Frank fuming mad and throwing my little Petri dishes into the autoclave bag for incineration.

I did not dare move or ask what was wrong.

I could clearly see the green and white fuzzy stuff growing in the Petri dishes. There were dozens of them in this experiment, which had been going on for weeks.

Now, weeks of data were ruined.

Frank pitched one Petri dish past my face in frustration, passing so close to me it was as if I was at home plate and the pitcher threw a classic brushback pitch to rattle my nerves. Frank was certain I had rushed the feeding (addition of growth factors) to play squash.

He was at least partially right.

I simply hated counting colonies and preparing those agar plates without an air bubble.

Recognizing my failure, I said simply, "Okay, I'll take the project with the deadly retrovirus."

Clearly, I wouldn't survive long in science doing colony assays. I decided to fall back on my viral purification skills.

However, this choice would bring me into direct conflict with the then reigning king of corrupt science, Robert Gallo, and his crown prince, Anthony Fauci.

* * *

Frank and two other scientists from NIAID had been working on confirming the work of the French researchers, Luc Montagnier and Françoise Barré-Sinoussi, confirming their isolation of the HIV retrovirus. At the time the virus was called LAV (lymphadenopathy virus) and AIDS was called GRID, which stood for Gay-Related Immune Deficiency.

Frank had trained me in the lab culture procedures to do the work using primary cells and the Hut-78 cell line, just as he'd trained Bernie Poiesz, prior to their discovery of HTLV-1. We had a cell line growing the virus for characterization and confirmation. Frank wrote the paper. As a technician in those days, I was not allowed to be a coauthor, although I had certainly read and made certain the experimental procedures were exactly as I had done them. That has changed in the ensuing years, and technicians can be credited in the publication as a coauthor.

However, somehow Gallo found out and he and Fauci called Frank's office when he was out of town. Since I was his only employee, I answered the phone and took care of the mail when Frank was out of town.

That day, in the late summer of 1983, I answered Frank's phone, to find Anthony Fauci and Robert Gallo on the line, very angry. They were looking for the manuscript describing the methods and virus isolates from confirmatory studies we had done concerning the lymphadenopathy virus, LAV, later called HIV, isolated from patients suffering from GRID.

I replied that Frank was out of town and since I was not a coauthor, I lacked the authority to give it to them.

Of course, they threated to have me fired, wreck my career, etc., and I hung up on them.

But just as I thought I'd blown it with Frank in my initial interview when I said it didn't matter who authored the papers, it was all about the data, Frank appreciated why I'd refused to hand over his paper. I wasn't just defending Frank from Gallo and Fauci.

I was defending a principle.

When I previously told this story, I'd said that Bruce Chabner, then head of the NCI, was on the phone as well. Frank remembers it differently, with me telling him it was just Gallo and Fauci. However, when Frank returned from Europe, Chabner did pressure Frank to give a copy of his paper and the important viral isolates to Gallo. Frank refused to do so. They held up Frank's paper until Gallo published his own isolations.

That paper won Gallo a second Lasker Prize in 1986.

Although it was in accord with the rules of the time, I believe Frank Ruscetti and Bernie Poiesz should have shared credit with Gallo in the Lasker Prize of 1982, for the isolation of the HTLV-1 virus, the first identified disease-causing human retrovirus. As Candace Pert and others found out later, the prizes go to laboratory heads, no matter how uninvolved they were. Almost all laboratory heads support this system because they want their prizes.

In my opinion, the awarding of the 1986 Lasker Prize to Gallo (for which was originally claimed to be the "discovery of HIV," but later amended to be a "confirmation" of Montagnier and Barré-Sinoussi's work) was a complete travesty. Gallo and Fauci pulled off a heist of Frank's paper, held it hostage for months, then Gallo published his own paper, using our methods.

The 1986 Lasker Prize should have gone to Frank Ruscetti and Bernie Poiesz.

* * *

By contrast, Luc Montagnier is a man of integrity.

That is why my coauthor, Kent, continually hounded me to send Montagnier a copy of our book, *Plague of Corruption*, prior to publication, to see if he would give us an endorsement. We edited his endorsement slightly (his English was a little fractured), but it is on all the hundreds of thousands of copies worldwide and reads, "The rampant corruption . . . hides from the public scientific truths which might go against corporate economic interests." I don't think you could get a more ringing endorsement.

And after the book was published, Kent exchanged several emails with Montagnier's personal secretary, Suzanne McDonnell, who expressed what I believe was Montagnier's personal excitement over the tremendous public response to the book.

To Francis Ruscetti, Congratulations! *Louis W. Sullivan, M.D.*

Frank receiving the Distinguished Service Medal from the National Institutes of Health in 1992, the highest honor a scientist can be given by the agency.

Everywhere we went it seemed people wanted books.

When *Plague of Corruption* hit #1 of all books on Amazon, Kent had to memorialize the occasion.

Judy and Kent spoke at an anti-lockdown protest at the California State Capitol in June 2020.

Robert F. Kennedy Jr. joined us at the protest and took this picture with our book, for which he wrote a wonderful foreword.

David has always been my rock!

Mikki Willis really helped us with his twenty-six minute interview of me which reached a BILLION views, even under relentless attack from the social media giants.

Since 1983, Frank and I have talked almost every day, usually arguing like cats and dogs over what the data mean, but never ignoring the data. (Courtesy of Taylor Mohr Photography)

Ever since I met and fell in love with David in 1999, Southern California is home for me. (Courtesy of Taylor Mohr Photography)

Frank and I are quite the pair. Once at a conference some people saw us arguing very heatedly about some data and became concerned. "Don't worry," said somebody who knew us well, "That's just the way Judy and Frank are. They really love each other." (Courtesy of Taylor Mohr Photography)

The churches have been the most welcoming of all groups, understanding the evil at the heart of this corruption of science.

Kent and I onstage at Calvary Church in San Jose, California, a center of the resistance against the insane COVID-19 lockdowns.

Over ten years Kent and I have written four books to try and end this corruption of science. Nobody is Superman in this fight, but together we can be a team of Avengers and restore faith in science.

Dear Kent:

We applaud your great, deserved success for you and Judy for PLAGUE OF CORRUPTION.

Perhaps Professeur's complimentary copy got lost in transit, but he has never received it. We would be so pleased to have the book resent.

Mme Suzanne McDonnel.

Bad Kent! Bad Kent! Next time you must remember to send a copy of our book to the Nobel Prize winner who endorsed it. Yeah, as if it wasn't a crazy time with everything going on with COVID-19 and the relentless media attacks. But Kent quickly recovered, writing:

Suzanne:

Thank you so much for the address. I tried to find it in my emails when the book came out and could not. I will be sending you two copies, one for you and one for Dr. Montagnier.

All the best,
Kent Heckenlively

Suzanne wrote back:

Thank you so much!

What you are doing, you, Judith, Robert, Mary, Lyn, Professeur, is God's work. May you, your efforts, your families be blessed.

xox Suzanne

Now when Kent showed that email to the documentary filmmakers and said, "Hey, we've got a Nobel Prize winner saying we're doing God's work," they were cautious.

"Maybe it's just his assistant who's telling you you're doing God's work," they replied.

Even at the age of eighty-eight, Montagnier continues to contribute to science, most notably with his recent thoughts on the origin of COVID-19, which mirror some of my own conclusions.

This is an article from *The Week* on April 19, 2020, just as some of the genetic information about the coronavirus was being published in scientific journals:

> French virologist and medicine Nobel laureate Luc Montagnier has made explosive revelations regarding the origin of the coronavirus, saying that the deadly virus was manufactured in a laboratory in China's Wuhan.
>
> Montagnier's claims come at a time when the US has alleged the possibility of the virus originating in a lab in China. The theory that COVID-19 was created in a Chinese lab and "leaked" out to the world has been making the rounds since its outbreak in December 2019. President Donald Trump also fired a fresh salvo when he warned China of consequences if it was "knowingly responsible" for the virus. China has refuted these allegations.[7]

Even though Fauci, the CDC, and the public health system have maintained for eighteen months that this was not the case, Fauci has recently turned around on this question and said the possibility should be investigated. The fact that China has denied these allegations and stonewalled any investigation makes me wonder.

Montagnier was a smart man, Frank always respected him, and every time I've interacted with him, he's been nothing but intelligent and warm. Before I was eviscerated over XMRV, I gave a talk at one meeting on March 3, 2011 right after Luc. Before I started to speak, he gave me the thumbs up!

His opinion about the coronavirus and COVID-19 mirrors my own thoughts from early in the outbreak. The earliest red flag to me was when the media gave us only two options, first, that it was a naturally occurring virus, or second, it was a "Chinese bioweapon!"

Wasn't there a third, logical possibility?

Escape of a highly pathogenic organism from a lab? That seems reasonable, and in my mind, the most likely possibility.

It's supported by publications in which the Chinese thank Anthony Fauci's NIAID and the US Army Medical Research Institute of Infectious Diseases (USAMRIID) for the Vero E6 monkey kidney cell line. That's why I say SARS-CoV-2 is most likely a monkey virus and NOT a bat virus. I worked in that USAMRIID facility in the 1990s using the Vero E6 monkey kidney cell line to grow different strains of Ebola, while a staff scientist in Frank's lab.

On April 8, 2020, I was featured in a video put out by the *Epoch Times* with reporter Joshua Philips commenting on this issue. I am convinced there are many parts of the SARS-CoV-2 story which still remain hidden. For example, the Vero E6 monkey kidney cell line that USAMRIID gave to the Chinese was NOT the only source of that cell line in the world. The Chinese could easily have purchased the Vero E6 monkey kidney cell line from the American Type Culture Collection without the involvement of the American government.

I strongly believe that our government gave the Chinese the Vero E6 monkey kidney cell line from Fort Detrick where I worked, that was also contaminated with Ebola virus and many other viruses. It is difficult for me to believe this was accidental. This suggests to me that some nefarious plan between our intelligence agencies and the communist party of China is behind this plandemic. These people are not stupid. They are smart and they are likely evil, in that they know how to distract the public from looking at what may be lurking behind the curtain.

But why didn't the mainstream media want us to consider even the possibility that this was a lab leak? It only got more interesting when the genetic code of the virus was revealed. It didn't look natural. It looked assembled. Montagnier and I saw the same clues:

> Montagnier alleged the presence of elements of HIV and germ of malaria in the genome of coronavirus is "highly suspect" and it "could not have arisen naturally". The French researcher also alleged an "industrial accident" to have taken place in the Wuhan National Biosafety Laboratory, which specializes in coronaviruses since the 2000s.[8]

I remind you that the precursor to HIV came from primates. What are sequences from a monkey virus doing in a bat virus? Did some bat from China have a torrid affair with an African monkey? I think not.

Why are we not allowed to ask the most basic questions about this virus and its origins? China doesn't want us to know, and I think there are elements of the US government which do not want us to know as well.

* * *

Let's talk about another misogynistic gatekeeper in science, Dr. John Coffin, an initial supporter, who became my most fervent critic and nemesis.

What caused the change?

On October 8, 2009, Coffin and his fellow collaborator, Jonathon Stoye, wrote an accompanying opinion piece to our *Science* article entitled, "A New Virus for Old Diseases?"[9] Coffin clearly reviewed our manuscript and got the NCI to allocate somewhere between eighty and a hundred thousand dollars to develop reagents to test for the virus BEFORE our October 2009 publication in *Science*.

Now, prior to the XMRV issue I'd never met John Coffin. Both Coffin and Frank received their doctorates in 1972, but Coffin always seemed to act toward Frank like a know-it-all older brother. When Frank had told Coffin early in his career that he was trying to isolate a disease-causing human retrovirus, Coffin told him, "Don't bother. They don't exist."

Frank then isolated the first identified disease-causing human retrovirus, HTLV-1, and then a few years later HIV burst on the scene, killing millions, and proving to everyone on the planet that retroviruses could cause disease.

Dr. Coffin wasn't setting himself up to have a good track record with predictions.

Although I suspected it at the time, but couldn't prove it, I believed John Coffin was one of the three anonymous reviewers of our 2009 XMRV paper in the journal, *Science*, which kicked off everything.

When Coffin was interviewed by documentary film director, Michael Mazzola, he admitted he was one of the three anonymous reviewers of the 2009 *Science* paper. Scientific ethics requires that if you are an anonymous reviewer on a paper, you refrain from public comment.

Bad move for Coffin, because that means his subsequent work, stage-managing the XMRV issue in the scientific community, broke the rules of scientific ethics.

Not that I expect him to ever be prosecuted.

The good ole' boys of science protect their own, and Coffin is one of them.

I just thought it's something I should mention for the record.

In addition to his position at Tufts University, Coffin in 2009 held a curious title, "Special Advisor to Director," for HIV drug resistance at the NCI.[10] This was an unusual arrangement because in 1995 Coffin had been hired by an Inter-Personal Agreement (IPA) from Tufts University to run the HIV drug resistance program. An IPA is only supposed to run for three years, but Coffin had remained with the HIV drug resistance program for ten years.

This was done by simply creating a new title for him.

When you're part of the group that makes the rules, it's easy to find a way to break them.

Along with Dr. Stuart Le Grice, Coffin organized the July 22, 2009 "Public Health Implications of XMRV" meeting at the Center for Cancer Research Division of the National Cancer Institute to discuss my findings, as well as those of others on the health implications of XMRV.

This was the confidential summary of the meeting and we covered it at length in chapter eight of our first book, *Plague*:

> In 2006, the human retrovirus XMRV (xenotropic murine leukemia virus-related virus) was identified and reported to be associated with certain cases of prostate cancer. Although the public health limitations of this finding were not immediately clear, a series of presentations at the most recent Cold Springs Harbor Laboratory meeting on Retroviruses provided additional support for this linkage and suggested that the number of individuals infected with XMRV is significant to be a cause for public concern. In view of these developments, it was deemed appropriate for NCI [National Cancer Institute] to convene a small group of intramural and extramural scientists and clinicians with expertise in this area to provide the NCI leadership with recommendations on future directions.[11]

Frank was again taking the lead in a critical investigation regarding public health. Coffin probably didn't want to be caught flat-footed again, the way he had with HTLV-1 and HIV.

For better or worse, Coffin was going to control XMRV.

A couple critical things stood out from this summary.

First, XMRV had first been found in association with prostate cancer in 2006 by a PCR-based technology, spearheaded by Dr. Joseph DeRisi of the University of California, San Francisco. Never had it been isolated and associated with ANY disease until my team isolated it from patients with myalgic encephalomyelitis/chronic fatigue syndrome (ME/CFS) and mantle cell lymphoma in 2009. ME/CFS occurs mainly in women. Anthony Fauci and others dismissed the disease thirty years earlier by declaring the patients were simply crazy or suffering from conversion disorder. In others words, they were mainly career women who just couldn't handle the stress of the male-dominated corporate world. One day these women woke up and simply didn't have the stamina to survive in a man's world. It's incredible to me, even though I lived through it, that these misogynistic gatekeepers of science have been allowed to keep their positions over the last three decades.

Further, others were finding troubling results like we did in association with various conditions. We did not present at the 2009 Cold Springs Harbor conference, because our manuscript was submitted May 4, 2009, and was confidential. But John Coffin certainly knew about it.

Finally, this meeting took place at the end of July 2009, while our results wouldn't be published in *Science* until October 2009. The only explanation that makes sense to me is the public health to have a good battle plan in hand when this was revealed to the public.

They didn't want to start a "panic."

This is a summary of Coffin's talk at the invitation-only July 22, 2009 meeting to discuss XMRV:

> Dr. Coffin discussed the properties of XMRV and its relationship to xeno-tropic murine leukemia virus (XMLV). He pointed out that different XMRV isolates are very closely related to, yet distinct from, endogenous proviruses found in the sequenced genome of an inbred mouse.
>
> These observations alleviate concerns of laboratory contamination with virus or DNA from laboratory mice but are consistent with very recent, per-haps ongoing transmission from a wild-mouse reservoir into the human pop-ulation. However, it is also possible that another animal species has been the vector for zoonotic infection.[12]

Let's break it down. There is a well-known mouse virus, XMLV (xenotropic murine leukemia virus), but while XMRV is very similar, it is "distinct" from XMLV. The genetic evidence was "consistent with very recent, per-haps ongoing transmission from a wild-mouse reservoir into the human population."

My coauthor, Kent, probably in about 2010, had asked me, "Judy, where was the first outbreak of chronic fatigue syndrome?" I told him I didn't know.

He did some research and came back with an answer.

It took place in 1934–1935 among 198 doctors and nurses at Los Angeles County Hospital, who were working during a polio epidemic.

But curiously, the disease did not appear among any of the patients.

What was different about the medical staff?

It turns out they'd been given an experimental polio vaccine that was grown in mouse brain tissue, along with an immune system booster preserved in thimerosal, a recently developed mercury derivative.

The vaccine was developed by Dr. Maurice Brodie, a young Canadian physician who worked at both the New York City Health Department and New York University.[13] Money for the vaccine came from the Warm Springs Foundation, created by President Franklin Roosevelt, as well as the Rockefeller Foundation.[14]

You can call me a cynic, but it seems if you want to start a cover-up, it would be good to have on your side an academic institution, the president of the United States, and a tremendously rich health foundation. And if it was covered up in the 1930s, others who followed would be expected to conceal the information from further generations.

In the early days, before everything went terribly wrong with the 198 doctors and nurses at Los Angeles County Hospital, here's what Dr. Maurice Brodie wrote:

> From these experiments, it appears that of all the ordinary laboratory animals, the mouse should be the best in attempting to produce poliomyelitis, for the virus survives in its brain for a longer time than in that of the guinea pig, rabbit, or rat.[15]

What most people don't realize about the viruses that are used in vaccines is you've got to first grow the virus and it's often very difficult to find a tissue in which the virus will grow to the extent needed. Brodie went on in the article to describe how he used the mouse brain tissue.

> The passage material was a mixture of 1 part of active virus and 4 parts of a suspension of mouse brain and brain stem. The material from the 24th and 45th passage injected into a series of mice in multiple inoculations, produced no effect in the mice nor did the virus in either of these passages survive for a longer time than in the preliminary experiment, when it was demonstrated in the mouse brain, 3 days but not 5 days after intracerebral injection.[16]

This vaccine was given to "300 nurses and physicians" from the Los Angeles County Hospital, as detailed in an article by Maurice Brodie and William Park.[17]

But when later researchers looked at the question of whether the mouse biological material was a good medium in which to grow polio virus, then inject into the bloodstream of a human being, they had concerns. The yellow fever vaccine also used mouse brain tissue, which was then injected into monkeys prior to injecting it into human beings. (As an aside, Dr. Amy Yasko, a leading autism doctor, has noted that some of the very earliest cases of autism were among the children of tropical disease researchers, the very families who would be expected to receive a yellow fever vaccination.)

A presentation made by the World Health Organization in 1953 specifically addressed the question of whether the mouse brain tissue used in the yellow fever vaccine presented unexpected dangers:

> [T]wo main objections to this vaccine have been voiced, because of the possibility that: (i) the mouse brains employed in its preparation may be contaminated with a virus pathogenic for man although latent in mice . . . or may be the cause of demyelinating encephalomyelitis: (ii) the use, as an antigen, of a virus with enhanced neurotropic properties may be followed by serious reactions involving the central nervous systems.[18]

Kent reviewed these materials and had many questions. "There can be viruses which do nothing in the animal, but when injected into human beings will wake up and cause damage?"

"Yes," I replied. That was the reason the most dangerous viruses are those which jump from animals into humans.

"These viruses can cause inflammation and demyelination of the nerves?" Kent asked.

"Yes."

"If the viruses don't wake up, or cause inflammation and demyelination of the nerves, it might still cause a serious reaction involving the central nervous system?"

"Exactly," I replied.

The scientists weren't too far behind Kent's excellent questions and they published an article in January 2011 in *Frontiers in Microbiology*, which came to many of the same conclusions.

And despite the fact that mice or mouse tissue may have been removed from general usage, the fact is it was already circulating in the human population, and also in cell cultures which were still being used.

In other words, it was like closing the barn door, after the horse had already run away.

Perhaps it was these realizations, or the presentations at the July 22, 2009 meeting, that resulted in Dr. John Coffin reportedly leaning over to one of the participants and saying, "Oh my God? You mean all those sequences we saw in the 1980s were real?"[19]

* * *

We may have stumbled into this problem in 2009, but this is likely to have been a problem from the 1930s, and it's only been the advances in technology which have allowed us to identify the problem. Maybe those sequences many scientists saw in the 1980s weren't "junk" as thought at the time, but the first indication we had pursued a dangerous course for humanity.

Was it in 2010, when I started talking about autism and its possible linkage to this virus and vaccinations, that Coffin decided to put all his available energy into killing this line of inquiry? It's difficult to come to any other conclusion, and when combined with what I consider to be his raging hatred of women in science, it made for a very toxic brew. This is a direct quote from Coffin which he provided to Jon Cohen of *Science* magazine for his "False Positive" story of September 23, 2011, which tried to put the XMRV genie back in the bottle:

"I began comparing Judy Mikovits to Joan of Arc. The scientists will burn her at the stake, but her faithful following will have her canonized."[20]

Perhaps John Coffin thinks it's appropriate to burn women at the stake for challenging powerful men. But that's not likely to be a widely shared view among the population in the twenty-first century. Maybe that "faithful following" of which he speaks is the honest women, men, and children who are suffering from this plague of corruption.

I'm not even sure the burning of Joan of Arc at the stake was a popular opinion in the fifteenth century when she was murdered. Her death provoked such an outrage, revealing corruption within the French royalty and the Catholic Church which had condemned her, and eventually leading to the English being driven from France. This is what *Science* magazine wrote about the autism issue:

"You don't talk about autism in the US, it's too politically charged," Mikovits claims Coffin told him. She believes Coffin turned against her that very day. Coffin confirms that he was upset that Lombardi [my colleague at the time] presented such preliminary data on such a fraught topic, but says, "I did not 'turn against' Judy at that or any other point."[21]

If this was a court of law, my testimony would be accepted as factual. Coffin admits he was "upset" that I was talking about such a "fraught topic."

And how can he claim with a straight face that he did not "turn against" me when he invited scientists to "burn" me at the stake?

What does Coffin say in explanation to my research? Here's what he told journalist Hillary Johnson in a 2013 *Discover* magazine article:

> [A] government consultant named John Coffin, had his own theory: XMRV had been accidentally created in a lab at Case Western University in Cleveland sometime between 1993 and 1996. Coffin described how lab workers there had transplanted human prostate tumor cells into an immune-deficient lab mouse, a common procedure for procuring a colony of cells, or a human cell line, for further study.
>
> When the human cells failed to thrive, lab workers transplanted those same cells into another mouse. Coffin and his collaborator, Vinay Pathak, suggested that with each passage, the human cells acquired genetic portions of a murine leukemia virus, which then merged to form a new virus, a hybrid of the parent sequences.[22]

The public probably doesn't realize it's a "common procedure," to use animal tissue to grow human cancer cells, as well as human viruses, just as Dr. Maurice Brodie did with the polio virus vaccine and mouse tissue in the 1930s.

It's called mouse xenografts and they did it every single day with patient-derived tumors in order to test drug efficacy in actual patient tumors. There is a 2021 paper in *Nature* with the title, "Presence of Complete Murine Viral Genome Sequences in Patient-Derived Xenografts." The abstract reads in part:

> Patient-derived xenografts are crucial for drug development but their use is challenged by issues such as murine [mouse] viral infection. We evaluate the scope of viral infection and its impact on patient-derived xenografts by taking an unbiased data-driven approach to analyze the extensive presence of murine viral sequences covering entire genomes in patient-derived xenografts. The existence of viral sequences inside tumor cells is further confirmed by single cell sequencing data. Extensive chimeric reads containing both viral and human sequences are also observed. Furthermore, we find significantly changed expression levels of many cancer, immune, and drug metabolism-related genes in samples with high virus load.[23]

I believe the use of animal tissue and aborted human fetal tissue in the man-ufacture of vaccines and other medications, commonly called "biologics," to be at the heart of this debate, and the one thing Big Pharma doesn't want the public to understand.

Thank you, Dr. Coffin, for making that crystal clear to people.

However, I took great exception to the conclusions Coffin drew from this information and said so in what seems to have been the final talk of my science career in September 2011.

Despite being jailed and bankrupted, along with the deliberate misrep-resentation of the Lipkin study in 2012, there were still courageous doctors and journalists with integrity like Hillary Johnson who wrote about the corruption in the brilliant foreword to our first book, *Plague*. In the last line of that foreword, Hillary quoted a career NIH official who, when asked why government officials don't investigate the very real scientific questions raised by this issue, replied that "They hate you." The scientist who said that was Frank Ruscetti, forced into retirement in 2013. We can't have honest scientists telling the truth, can we?

Here's how Coffin continued to try and bury XMRV, as detailed by Hillary Johnson:

> Obviously, if XMRV had been inadvertently manufactured as the result of lab experiments in the mid-1990s, it could not be causing disease in CFS or prostate cancer patients who had fallen ill in the 1980s or early 1980s. XMRV was "dead," Coffin announced in April 2011. He advised CFS researchers to "move on."[24]

Why then did Coffin tell everyone to move on, even as his lab continued to publish data that supported the idea that these animal viruses and their sequences had contaminated cell lines used in the manufacture of vaccines since at least the 1960s? Several slides in my talk in September 2013 are the cover-up of information from the lab of Gary Owens, first shown at the Cleveland Clinic on November 10, 2009, showing a variant of XMRV which he named "XMRV-2." However, a subsequent publication called it B4RV, to cover up the reality that our team was right.

Slide 30 of that presentation shows a *Journal of Virology* paper, coau-thored by Coffin, which shows several strains of XMRV and herpes virus contaminating a commonly used cell line.[25]

Therefore, I take great exception to Coffin's call for people to move on from XMRV, which he believes was created between 1993 and 1996. As his own paper showed, the mixing of human tissue and animal tissue in cell culture will often generate new, replication competent retroviruses (and coronaviruses?) in only two weeks.

That is, in every single new lot of vaccines.

So, what about the fact we've been doing this very thing for more than a hundred years in vaccine development? Were we really expected to believe it only happened once? And now, in 2021, papers are being published trying to convince people that the viruses in these human xenografts are jumping from the human to the mouse cells, but not the other way around. This is extremely unlikely.

In Coffin's view, XMRV had to be what I dubbed "the immaculate recombination," meaning that this type of virus was generated in only one time at one place in history, despite the fact we've been conducting similar experiments for more than a hundred years.

However, you don't have to take my word for it, as we have strong evidence of XMRV existing before Coffin's 1993 to 1996 time period:

> Shyh-Ching Lo, a pathologist and director of the FDA's Tissue Microbiology Laboratory, discovered that he possessed 37 unopened vials of whole blood and plasma from CFS sufferers, the pristine cryopreserved remains of experiments undertaken in the early 1990s. To his surprise, Lo found four different MLV-related gene sequences in 32 of the 37 patient samples. [Approximately 86 percent, significantly higher than the 67 percent we reported in *Science*.][26]

This is probably about as clear as it gets in science. Yet, Coffin persisted in this unscientific explanation at the April 2011 State of the Knowledge workshop. Harvey Alter, the 2020 Nobel Prize winner in Medicine and senior author on Lo's 2010 study, took me aside and privately said, "This makes no sense at all. Do you understand this immaculate recombination?" Before I could answer, Alter continued, "I can certainly tell you there was absolutely no evidence of contamination in Lo's study."

Coffin suggested XMRV was generated in a lab sometime between 1993 and 1996, eventually contaminating lab samples worldwide. Shyh-Ching Lo had samples taken from actual patients in the early 1990s, before Coffin's timeline. The findings were consistent with our research. In addition, they found some of those same people from whom the samples had been taken in the early 1990s, drew fresh samples from them in 2010, and were again able to find evidence of the virus.

This was a top-notch investigation, and their findings should have been the confirmation of our work, resolving any lingering doubt and turning the research community toward finding solutions.

> Although the gene sequences were not identical to the Mikovits-Ruscetti XMRV gene sequence reported in *Science*, they were so close Lo believed he had found genetic variants of a single MLV-like virus species that likely included XMRV. Lo was encouraged by the variants because retroviruses are extremely mutable pathogens that change their gene sequences again and again in response to immune system efforts to kill them. "This is what we expect," Lo said at the time.[27]

Our discovery of XMRV was confirmed by leading scientists and what we found was consistent with known science about viruses. But Coffin just wanted to muddy the waters.

And what happened to Dr. Maurice Brodie, whose mouse-derived experimental polio vaccine probably caused the first outbreak of chronic fatigue syndrome (ME/CFS) in 1934–1935 among 198 doctors and nurses at Los Angeles County Hospital?

Despite the likely help of New York University, President Franklin Roosevelt's polio foundation, and the Rockefeller Institute's financial resources in helping cover-up the outbreak, Maurice Brodie died in May 1939 at the age of thirty-six, in what was a suspected suicide.[28]

* * *

Without a doubt, another New York criminal is Columbia University's Ian Lipkin, whom the *New York Times* proclaimed in a 2010 article spends his afternoons, "prowling his empire of viruses."[29] The writer of the article, Carl Zimmer, described Lipkin in breathless prose more suited for an action hero trying to defuse a ticking time-bomb:

> Rather than wait for the elevator, Dr. Lipkin ran up and down the back stairs to move from floor to floor, leaning into the doorways of labs and glass-walled offices to get updates from a platoon of scientists.
>
> Gustavo Palacios was sequencing the genes from a new strain of Ebola virus found in a bat in Spain, a worrisome development, since the fatal virus has almost never been found outside Africa.

Nick Bexfield of the University of Cambridge had flown from England with a new hepatitis virus that had just broken out in British dogs.

. . . all told, members of Dr. Lipkin's team were working on 139 different virus projects. It was, in other words, a fairly typical day.[30]

I think of Ian Lipkin as a well-spoken assassin for Big Pharma.

Frank warned me about trusting Ian Lipkin, and I should have known the trouble that would be brought to me courtesy of Anthony Fauci, as the XMRV-associated disease nightmare was becoming more heated for the NIH.

In the end, the NIH's AIDS czar, Anthony Fauci, asked his friend Ian Lipkin, a neurologist and virus hunter at the Center for Infection and Immunity at Columbia University's Mailman School of Public Health, to settle the impasse.[31]

What does the data say about Lipkin's discoveries? Actually, Lipkin spends most of his time debunking and repackaging the discoveries of others, and then selling these lies to the public in order to continue the eugenics and depopulation agenda of his biggest funders, Anthony Fauci and Bill Gates.

What have others had to say about the quality of Ian Lipkin's work when he wades into controversial research? Do they say good things about him?

They do not.

My coauthor, Kent, conducted a long interview with Dr. Andrew Wakefield in 2016 about the quality of Lipkin's work, trying to confirm or refute an association between the measles virus, presumably from the MMR (measles-mumps-rubella) shot, and the development of gastrointestinal problems and autism. This is what Wakefield said about the quality of Lipkin's work:

The greatest concern is that when we looked for the virus throughout the bowel, the large intestine, and the terminal ileum with its lymphoid tissue, it was only in this hugely swollen lymphoid tissue that we found the virus. We did not find it anywhere in the colon.

Now, Lipkin got his biopsies from Tim Buie, and Tim, bless his heart, was not very good at getting into the ileum, which is technically challenging to many gastroenterologists. I worked with guys who were very good at it. It wasn't me. I didn't do it. But they were very good.

So we were always able to get samples of this lymphoid tissue where we found the virus. When Buie did his studies with Lipkin provided tissues from the cecum (the large intestine), and the ileum, and he didn't say how many came from the respective sites. Well, all of them that came from the cecum would have been negative based on our earlier studies. It was most definitely not a replication study.

And what slayed me was Lipkin's subsequent reporting in the news media that this was the final word and it ruled out the possibility that the MMR was causing autism. It didn't even go close to doing that.[32]

Let me put these comments into perspective for you. Wakefield only found the measles virus in the swollen lymphoid tissue of the ileum.

Lipkin had Tim Buie take many samples, but didn't note where or even if they came from the lymphoid tissue of the ileum. We're just supposed to accept his word that SOME of the samples came from that region.

It's like a student telling his teacher he did his homework, but he just can't show it to the teacher. But he still wants the credit.

That is not good science.

And with that impeccable logic, Lipkin declared we should not do any further investigation into whether the MMR vaccine was linked to the development of gastrointestinal problems and autism.

* * *

It's important to understand the approach we used in our initial investigation into XMRV.

We hypothesized that the virus would only be intermittently in the blood, and in only the sickest patients. It's not the presence of the provirus in the genome, but the expression of the virus, that generates the cytokine storm, which is the real disease.

We published this data more than a year after the *Science* paper.[33] Not because we didn't have the data, but because of the misogynistic gatekeepers of science. The so-called peer-review process is little more than competitor regulation by the gatekeepers like John Coffin, Ian Lipkin, Robert Gallo, Anthony Fauci, and Harold Varmus, to control the message and prevent the publication of relevant data which can shift current understandings. That's why we not only identified the sickest patients, but also took blood at regular intervals over several months and tested each sample, using multiple assays, including the addition of 5-azactydine, the demethylating agent, to activate the expression of latent XMRVs.

If this was a virus which was easily detected, it would already have been done. I could go into a lot more of the science, but I think that's sufficient for you to understand why Lipkin's multi-center study was fraudulent collusion by John Coffin, Ian Lipkin, and Anthony Fauci. Coffin had no business being at the November 4, 2010 study design meeting for the study Francis Collins commissioned, and that Anthony Fauci assigned to Ian Lipkin. Nor did Simone Glynn, Nancy Klimas, or Suzanne Vernon. All had conflicts of interest and decades of scientific misconduct towards the ME/CFS population. I had no idea that two years later there would be twenty-five authors on a paper that should have had no more than five or six.

Here are the conditions that were specifically excluded from the Lipkin multi-center study:

> Potential CFS/ME and control subjects were excluded for the following confounding medical conditions: serologic evidence of infection with human immunodeficiency virus, hepatitis B virus, hepatitis C virus, *Treponema pallidum*, or *B. burgdorferi*; medical or psychiatric illness that might be associated with fatigue; or abnormal serum chemistries or thyroid function tests.[34]

These exclusions are insane, and I fought as hard against them that day as I do now. The condition is called chronic fatigue syndrome and you exclude a "medical or psychiatric illness that might be associated with fatigue?" These patients are often unable to function for days or weeks, and you don't expect them to have "abnormal serum characteristics?"

Dr. Paul Cheney, one of the leading experts on ME/CFS, was particularly concerned about the Lipkin study's exclusion of those with abnormal thyroid function tests:

> Thyroiditis is fairly common in this illness. And there was a recent report, I believe out of the UK, which said that 82% of people complaining of chronic fatigue had evidence of thyroiditis. Another paper showed that one of the tissue foci for HHV-6A [a herpes virus suspected in chronic fatigue syndrome] is the thyroid. Jim Jones, way back when [1980s] in describing the first cases of CFS, said 85% of them had auto-antibodies against the thyroid. So, by definition, 85 percent of ME/CFS patients have thyroiditis.[35]

Cheney went on to say that the exclusion of patients with "medical or psychiatric illness that might be associated with fatigue," was a "giant head-scratcher."[36] In summation he said, "This is a medical condition associated with fatigue. So, they're trying to exclude the very condition they're supposed to be looking for."[37]

You may ask why did I stay, and if I had so much importance, why wasn't the study done correctly? We told much of this story in our first two books, *Plague* and *Plague of Corruption*, but I'll summarize it here.

In addition, the timeline of events was not clear to us as much of what occurred happened after I was jailed and while I was locked out of the literature and under a virtual gag order as an NIH-designated "fugitive from justice" for five years, unable to set foot on NIH property in Frederick, Maryland, or at Fort Detrick.

It's difficult to prove who authored this series of outrages, but I believe it had to come from the very top, which means Anthony Fauci.

First, the study was highly unusual in that when the study was unblinded it was done so first in a highly secret government conference call, led by Anthony Fauci, Harold Varmus, and Francis Collins. I know this because the security on the call was voice recognition technology which identified all the participants. I was in my car, somewhere on the 405 freeway in Los Angeles.

When I heard the verdict of no association, I cried.

Once again, this population would have no voice or treatment.

But then Lipkin said there was evidence that 6 percent of the controls had evidence of XMRV infection and my spirits instantly lifted. I thought *Thank you Jesus* as I said out loud, "That's still twenty-five million Americans." These were people who could be prevented from ever developing ME/CFS, autism, autoimmune disease, and cancer.

Next, if you go to the list of study authors, you'll see there are twenty-five listed authors, with John Coffin being one of them. If it's true that he was one of the anonymous reviewers of the *Science* paper, that's another ethical violation on his part. So, I was one of twenty-five authors, which meant my input counted for about 4 percent. Do you think the deck was stacked against me?

And what if I'd refused to lend my name to the paper? The story would have been that I wasn't willing to submit my work to rigorous analysis. And all the while, both Lipkin and Coffin were privately telling me, "Judy, we know there are retroviruses involved in this disease, but we just need to solve the VP-62 issue and get this behind us. We're going to do more research with you on those other viruses."

To my shame, I believed them. But as soon as the study was over, Lipkin, despite his promises to include us in future studies, never called or returned the calls we made to him.

* * *

In my opinion, the Lipkin study was nothing more than commissioned fraud, and Lipkin was rewarded to the tune of thirty-one million dollars by his good friend, Anthony Fauci, with the establishment in March 2014 of a Center for Research in Diagnostics and Discovery (CRDD), under the auspices of a new NAIAD program titled Centers of Excellence for Translational Research.

They claimed:

The CRDD brings together leading investigators in microbial and human genetics, engineering, microbial ecology and public health to develop insights into mechanisms of disease and methods for detecting infectious agents, characterizing microflora and identifying biomarkers that can be used to guide clinical management.[38]

And who else was part of this new institution? None other than Dr. Ralph Baric of the University of North Carolina, Chapel Hill. You may have heard that name recently. He was one of the top American researchers studying "gain of function" research in coronaviruses in collaboration with the so-called "bat lady of Wuhan," Dr. Shi Zhengili. From the *New York Times* on June 14, 2021:

Ralph Baric, a prominent University of North Carolina expert in coronaviruses who signed the open letter in *Science*, said that although a natural origin of the virus was likely, he supported a review of what level of biosafety precautions were taken in studying bat coronaviruses at the Wuhan Institute. Dr. Baric conducted N.I.H.-approved gain-of-function research at his lab at the University of North Carolina using information on viral genetic sequences provided by Dr. Shi.[39]

Getting the spin? Yes, Ralph Baric, Ian Lipkin's collaborator, was conducting dangerous "gain of function" research on bat coronaviruses, but it was "N.I.H. approved."

Are you feeling better, now?

And if you think I've been beating a dead horse in going after Robert Gallo, well, the *New York Times* uses him as a character witness in defense of Dr. Shi Zhengili.

> "She's a stellar scientist – extremely careful, with a rigorous work ethic," said Dr. Robert C. Gallo, director of the Institute of Human Virology at the University of Maryland School of Medicine.
>
> The Wuhan Institute of Virology employs nearly 300 people and is home to one of only two Chinese labs that have been given the highest security designation, Biosafety Level 4. Dr. Shi leads the institute's work on emerging infectious diseases and over the years, her group has collected over 10,000 bat samples from around China.
>
> Under China's centralized approach to scientific research, the institute answers to the Communist Party, which wants scientists to serve national goals. "Science has no borders, but scientists have a motherland," Xi Jinping, the country's leader, said in a speech to scientists last year.[40]

The disgraced Robert Gallo is now a character witness for a communist Chinese scientist who likely unleashed our greatest modern plague. Kind of makes everything we've been writing about Gallo pale in comparison, right?

Another one of Ian Lipkin's seven collaborators in his new institute, Dr. David Relman of Stanford University, seems to tacitly acknowledge that Dr. Shi and her colleagues may have killed more than half a million Americans, and many more worldwide, but doesn't want us to be unpleasant with the potential author of all this suffering and death:

> "This has nothing to do with fault or guilt," said David Relman, a microbi-
> ologist at Stanford University and coauthor of a recent letter in the journal
> *Science*, signed by 18 scientists, that called for a transparent investigation into
> all viable scenarios, including a lab leak. The letter urged labs and health
> agencies to open their records to the public.[41]

Isn't it refreshing to know that a Stanford professor is encouraging us to not be harsh with a scientist who may be responsible for the death of more than half a million Americans? Do you ever get the feeling that many of these people in so-called "public health" aren't really interested in hearing the public's questions about their health?

But I want to be fair and give the last word to Dr. Shi Zhengili. From the *New York Times* article:

> In an interview with *Science* magazine last July, she said that Mr. Trump owed
> her an apology for claiming the virus came her lab. On social media, she said
> people who raised similar questions should "shut your stinky mouths."[42]

So much for Gallo's claim she was a "stellar scientist" who was "extremely careful with a rigorous work ethic."

Still, I was looking for the silver lining in these dark clouds of working with Ian Lipkin and company and continued to believe it was there.

It's just that the scientific community and press wanted to ignore them.

For example, the paper noted they'd found evidence of XMRV in 6 percent of the control population and 6 percent of the patient population.[43]

By their exclusions, they'd made both populations essentially the same, or at the very least, they'd chosen the relatively healthy chronic fatigue syndrome patients whose immune systems were somewhat keeping the virus dormant.

With a current population of 320 million people, that means a little over nineteen million people (not the twenty-five million I'd initially thought) are currently infected with these viruses, and the expression of just the gamma-retroviral envelope alone is strongly associated with disease development.

But the leaders of our public health establishment won't tell you that.

* * *

But you might say, well, you're just a Lipkin critic.

I'm sure the people who know and love and have worked more intimately with Ian Lipkin paint a much brighter picture of the man. Probably no one has been closer to Ian Lipkin than his longtime collaborator, Dr. Mady Hornig, with whom he also had a long-term romantic relationship, a fact which was widely known.

In May 2017, Hornig, whom I hold equally responsible for the destruction of my and Dr. Andrew Wakefield's scientific reputations, sued Dr. Ian Lipkin and included details of their romantic relationship. As reported in *Science* magazine:

> In the lawsuit, filed on 15 May in the U.S. District Court for the Southern
> District of New York, Hornig alleges that Lipkin for years has discriminated

against her on the basis of her sex and created a hostile work environment, violating U.S. and New York civil rights laws. In particular, it alleges that Lipkin took credit for Hornig's work; diverted or misused funds, thus delaying publication of Hornig's research results; and improperly added himself as principal investigator to grants.[44]

Lipkin is alleged to have engaged in abusive behavior with his longtime collaborator, as well as engaging in some unethical and illegal acts. The *Science* article went into further detail regarding the allegations:

> The lawsuit alleged that since 2013, Lipkin has refused to allow Hornig to post about her own work on the center's website unless the posting included him; required her to get his permission before giving invited talks; routinely presented Hornig's work as his own in meetings with collaborators; blocked her from meetings with potential donors; and silenced her at meetings, "sometimes kicking Plaintiff on the shins under the table . . . or saying 'shut up, Mady' or 'shut the f**k up, Mady' at meetings attended by both Columbia and non-Columbia colleagues. He also, she alleges, has repeatedly refused to support her for full promotion to full professor, even while supporting a male colleague.[45]

What Mady describes is a situation I've seen far too often with powerful men in science. They hold the establishment line and any woman who threatens that, or their fragile male egos, should be on notice. You have no idea what the wrath of an enraged and powerful male scientist can do to your life. And what should be most concerning to everybody interested in science, or how research money for crippling conditions is spent, are these allegations by Hornig:

> Among the claims of misuse of funds, Hornig alleges Lipkin paid the salary of a researcher studying CFS/ME [chronic fatigue syndrome/myalgic encephalomyelitis] with money from the Simmons Autism Research Initiative, which was supposed to be dedicated to an autism study. The suit also claims Columbia had to return more than $53,000 to the National Institutes of Health (NIH) because Hornig refused to sign off on improper use of autism grants.[46]

These are serious allegations, and if they were not true, I'd expect Mady Hornig to be fired for bringing them.

However, if they were true, I'd expect Ian Lipkin to be fired.

And yet, both continue to work at Columbia University, so what can I conclude?

The system protects itself once again?

And there are even more recent allegations that Ian Lipkin may be either a witting or unwitting agent of influence for the communist party of China.

In an article published by Fox News on July 1, 2021 with the title, "Columbia Professor Who Thanked Fauci for Wuhan Lab Messaging Has Links to Chinese Communist Party Members," it was detailed how not only has Lipkin praised Fauci for discounting the Wuhan "lab leak" theory, but also his close association with the Chinese communists:

> In 2016, the Chinese government presented Lipkin with the International Science and Technology Cooperation Award, the country's highest honor to foreign scientists. The ceremony was presided by President Xinping.
>
> "I am deeply honored by this award," Lipkin said at the time. "It solidifies my relationship with dear friends and colleagues in the Chinese Academy of Science, Ministry of Science and Technology and the Ministry of Health, and with the people of China."
>
> In early January of this year [2021], Lipkin was presented with a medal at the Chinese Consulate of New York. China's central government, central Military Commission, and State Council provided the award.
>
> Lipkin also praised China's response and transparency early on in the coronavirus pandemic, despite indications that the Chinese government engaged in a cover-up.[47]

It might be one thing to be given an award by the Chinese in 2016. Maybe a certain naiveté was involved and we might excuse it.

But to accept an award from the Chinese communist government in January 2021, after a Chinese virus, which likely escaped from their lab, killed hundreds of thousands of American citizens as well as millions around the globe?

Something just doesn't seem right with this picture.

Just for the record, Frank and I have never received an award from the communist party of China.

* * *

The misogynistic gatekeepers in science wouldn't be complete without their "cleaner," the fix-it guy who cleans up evidence from the crime scene, arranges for disposal of the dead body, and gives everybody their alibis.

In our current situation, that person is none other than Harold Varmus. Here's how Varmus describes himself in his biography:

> I have also worked in significant leadership positions: as Director of the NIH
> [National Institutes of Health] from 1993 to 1999, as president of Memorial
> Sloan Kettering Cancer Center from 2000 to 2010, and as Director of the
> National Cancer Institute from 2010 until April 2015, when I became the
> Lewis Thomas University Professor at Weil Cornell Medicine. I am also a
> Senior Associate Member of the New York Genome Center and teach a course
> in "Science and Society" at the CUNY honors college.[48]

People have talked about downward mobility, but Varmus must win the prize for that, just as he won the Nobel Prize in 1989 for the discovery of retroviral oncogenes.

In plain English, that means retroviruses insert some of their genes into your genes, and those genes in turn cause cancer.

If you're not worried by now about retroviruses because of chronic fatigue syndrome or autism, you should probably add cancer to the mix.

Might that have anything to do with our sky-high rates of cancer?

Let's go through Harold's career as he describes it. From 1993 to 1999 he was director of the NIH, which is the most powerful health agency in the entire government. Dr. Anthony Fauci, as head NIAID, was under Harold Varmus in the chain of command. According to its website, the NIH:

> The NIH invests about $41.7* billion annually in medical research for the
> American people. More than 80 percent of NIH's funding is awarded for
> extramural research, largely through almost 50,000 competitive grants to
> more than 300,000 researchers at more than 2,500 universities, medical
> schools, and other research institutions in every state.[49]

The NIH is an enormous funder of medical research and Harold Varmus, as the head, oversaw all that for six years. The yearly salary for the director can be as high as $230,000 a year, depending on your political connections.

My guess? With a Nobel Prize under his belt, he got top dollar. As described by the NIH website:

> The Office of the Director is the central office at NIH for its 27 Institutes and Centers. The OD is responsible for setting policy for NIH and for planning, managing, and coordinating the programs and activities of all the NIH components. OD program offices include the Office of AIDS Research and the Office of Research on Women's Health, among others.[50]

What are some of those "27 Institutes and Centers?" Well, the NCI is one. So is Anthony Fauci's NIAID, which he's run since 1984. The National Institute of Aging is another one.

Varmus spent six years at the top of the medical research funding pyramid. From there he went on to run the Memorial Sloan-Kettering Cancer Center from 2000 to 2010. It was reported that the 2016 salary for the then director of Memorial Sloan Kettering Cancer Center, Craig Thompson, was $6.7 million.[51] I think we can safely assume Varmus's salary was somewhere in that range.

But then the XMRV problem came up, and there was the issue of Frank Ruscetti at the NCI. How could that problem be resolved?

Call in Harold Varmus to shut things down.

The current director of the NCI, Norman Sharpless, earns $375,000, just under the $384,625 earned by Anthony Fauci.[52] We're supposed to believe that the man who had been in charge of the government's twenty-seven medical institutes and centers, then headed up a wealthy cancer charity where he made several million dollars a year, was only too happy to go to a job at one of those institutes, and he'd only made a couple hundred thousand a year?

Prior to 2010, the NCI was run by a fine man named John Niederhuber, who supported Frank's work.

But with Varmus in the henhouse, that was about to change.

With Harold Varmus at the helm, he brought in a group usually consisting of John Coffin and Steve Hughes.

Frank recalled to me that at one point, Hughes said to him, "This isn't an inquisition."

"Sure looks like one," Frank grumbled back.

I believe it was Varmus who ordered Frank's premature retirement in 2013 and also gave the order that all XMRV samples and materials were to be destroyed.

* * *

As of June 1, 2021, here's where the main figures profiled in this chapter stand:

Robert Gallo, eighty-three years old, still works as director of the Institute of Human Virology at the University of Maryland, as well as consulting with numerous bio-tech companies.

Anthony Fauci, eighty years old, still runs NIAID and has been the main scientific spokesman for the government during the COVID-19 crisis.

John Coffin, eighty years old, still works at Tufts University.

Ian Lipkin, sixty-eight years old, and despite serious allegations brought against him by his longtime collaborator Mady Hornig, of sexism and financial misconduct, continues to work at Columbia University. His history of working with the newest technology was described in this way by *The Lancet* COVID-19 Commission:

> These advances have been critical in replacing culture-dependent methods of global health management by creating new criteria for disease causation and de-linking spurious associations between putative agents and diseases. Such examples include refuting the MMR vaccine having a role in autism and XMRV in ME/CFS. Lipkin has been at the forefront of outbreak response to many of the world's recent outbreaks, including West Nile Virus in NYC (1999), SARS in China (2003), MERS in Saudi Arabia (2012-16), Zika in the US (2016), encephalitis in India (2017), and COVID-19 (2020). He went to China in late January 2020 to consult with colleagues at the China CDC during the early assessment of the SARS-CoV-2 outbreak and.[53]

Harold Varmus, eighty-one years old, runs the Varmus Lab at Meyer Cancer Center at Cornell University in New York City.

Frank Ruscetti, seventy-seven years old, was forced to retire from the NCI in 2013.

I am sixty-three years old, exiled from science since 2011. The fact that the Vaccine Injury Compensation program owes Mikovits and Ruscetti Consulting almost a half million dollars, and not only have not paid us for almost five years of work, but have also filed defamatory judgments against us, prohibiting us from working in the program or with other victims, is beyond criminal.

When one looks at the wreckage of 2020, it's easy to imagine that at some point the villains all got together to create some master plan, and then when their objectives were accomplished, reunited to pat themselves on the back for a job well-done.

Never could I have imagined that our book, *Plague of Corruption*, would sell out the day after it was published on April 14, 2020 and become the runaway science bestseller of the year, dwarfing sales of all other virus-related books. The people were hearing a message the mainstream media didn't want to be broadcast. And the people were responding, waking from their decades-long slumber, and starting to question some of our long-established institutions.

The battle for humanity's future is not yet over.

Spoiler alert—God has already won!

CHAPTER FOUR

The Filmmakers, an Angel, and Gifts Straight from God

Science is not only compatible with spirituality; it is a profound source of spirituality.

—Carl Sagan

In 2018 my mother became ill.

I first appreciated the severity of the situation in May, when she called me to say she was profoundly fatigued and depressed. In 2015, she'd been diagnosed with cardiac fibrillation (an irregular heartbeat), which put her at increased risk of a stroke. This terrified my mom as her mother-in-law had suffered a severe stroke when we were small children and she'd observed the terrible damage it did to my grandmother. As long as I could remember, my mother was very health conscious, following all the guidelines for heart healthy, low-fat eating and watching her weight. Though she was never into athletics, she enjoyed daily walks with her beloved dog, Bailey.

The phone call was very disturbing.

I don't recall my mom ever using the word depressed to describe herself, but even more concerning was the lack of life in her voice. Something was very wrong. She said my sisters were encouraging her but she couldn't function through the fatigue.

I was in Cleveland, Ohio, speaking at a Health Freedom Conference, so thank God I had a lot of time that day to try to get to the bottom of it. Nothing about this was making any sense.

My mother finally confessed that in 2011, right after I'd been released from jail, she'd been scared by her doctor into getting the flu and Prevnar (pneumonia) shots and had become very ill for almost six months. In my opinion, her illness was clearly ME/CFS, otherwise known as chronic fatigue syndrome. This was the very disease I'd been studying for the past several years, and for which I'd gotten into such trouble because I suggested that vaccines could be a trigger for people already carrying the XMRV retrovirus. My family had been afraid to tell me as they thought I had enough to deal with and did not want to worry me further.

I immediately changed my travel plans and flew back to Frederick, Maryland.

My mother was diagnosed with congestive heart failure, which by August had become so severe she was hospitalized and near death with severe anemia. To my horror, her cardiologist had prescribed the blood thinner, Eliquis, in 2015, and had never checked her hemoglobin levels. She had repeatedly complained of bleeding hemorrhoids since starting the drug. I was at home watching television when a commercial for Eliquis came on, with the warning, "In rare cases, life-threatening bleeding disorders can occur." No wonder she was tired and had congestive heart failure.

She'd been bleeding internally for three years!

A blood transfusion saved her temporarily from becoming yet another statistic of the third leading cause of death in America: medical mistakes! She was terrified of receiving XMRV-infected blood, but I assured her that the Cerus INTERCEPTsystem had cleaned up the blood supply.

As we were nearing the completion of *Plague of Corruption*, I was passing the hours in a Frederick, Maryland hospital with her, reading her sections of the book and trying to convince her it wasn't her fault she'd succumbed to medical bullying and taken the vaccine. When a nurse came into the room, I made certain to read very loudly, and shared with every nurse how my mother's idiot cardiologist denied any connection of vaccines or Eliquis to my mother's condition. Of course, the doctor blamed the patient.

I did my best to heal my mother with supplements and cannabis formulations, and she was able to go home. In those dark days at the hospital, nobody believed she would ever return home, but that blessed day did come.

I returned to California with renewed vigor and an urgency to educate as many health-care professionals as possible about the dangers of untested vaccines as we worked on case after case in vaccine court, seeing injuries similar to those suffered by my mother. Robin, a good friend of mine, was dealing with her father, who had Parkinson's disease and was in a nursing home. Against the written instructions of his family, he was given the flu and Prevnar shots, which accelerated the progression of his disease. He died shortly thereafter.

Our family's relief that our mother was healing was short-lived, as weeks later she was hospitalized with a bowel obstruction and given a diagnosis of incurable cancer.

I flew back immediately and repeatedly asked to see the pathology report but was not provided with it. Weeks later, after she'd been bullied into considering chemotherapy, I received the pathology report which showed the tumor was largely benign, had a few cancer cells, and that her liver was completely clear. In fact, the tumor was likely caused by the bleeding hemorrhoid and Eliquis, as her TGF-beta was literally trying to tie a rope around the area and stop the bleeding.

This unconscionable cascade of medical malpractice resulted in my mother's death on January 22, 2019. They had killed my mother with their vaccines and Eliquis and then claimed it was my mother's fault for not exercising.

The murder of the elderly with the flu/Prevnar vaccines enraged me and I tried once again to get my whistleblower lawsuit out from under the seal of the Nevada Attorney General's Office and the Federal Bureau of Investigation (FBI), where it had been languishing for more than five years. Their failure to act was preventing the public from understanding the risks of vaccine contamination and how this contamination was leading to the explosion of cancer and neuro-immune diseases.

I called my attorney, David Follin, who had worked tirelessly, but to no avail, with the Nevada Attorney General's office. In fact, on March 16, 2017, we had flown to Nevada and presented all the evidence to an office full of government agents. More than two and a half years had passed since that meeting and the government had DONE NOTHING.

Follin inquired again in early 2019 but received no response.

Follin told my husband he would no longer take my money. The Nevada Attorney General's Office had made it clear to Follin that if I continued to pursue the case, my family would once again be at risk.

Finally, I consulted with paralegal Travis Middleton, who helped me file a criminal affidavit against the Nevada Attorney General's office and the FBI. As you can probably guess, we received no response.

I couldn't help wondering, *is this still the United States of America?*

* * *

One blessing in this nightmare is that I began to understand the critical role of TGF-beta as the main driver behind vaccine injury and the subsequent development of cancer and neuro-immune diseases. I received an invitation to give a talk at a stem cell meeting in Ft. Lauderdale, Florida, March 22–24, 2019, sponsored by the Academy of Regenerative Practices. Now this was a first. Of course, I called Frank for help. The title of my talk was "Myelopoiesis: Key Process in Neuronal Immune health modulated by Cannabinoids." I also began to more deeply appreciate the path God had laid out for me in my graduate school years with my study of TGF-beta, and my doctoral thesis on HIV latency in monocytes.

As I write this in 2021, it is even clearer to me that SARS-CoV-2 is not the cause of COVID-19, but that a large part of the answer lies with TGF-beta and DNA methylation. The scientific community has been looking at the wrong target, just like in HIV/AIDS.

The talk, as Frank had taught me, demanded intelligence from the audience. I was busy trying to memorize those all-important transitions so that people understood the thought process. The talk was explained at the level of the blood stem cell and TGF-beta a lot of what I learned from Drs. Chris Shaw and Stephanie Seneff and cases in Vaccine Court. I wasn't paying much attention to the lunchtime speaker as my talk was right after lunch. I stood next to the stage, a bit nervous, as the speaker, ranting about mercury toxins in dentistry, went over into my time. I knew I'd need every minute to explain these slides. As I was getting a microphone on, I couldn't help but notice a picture on Dr. Lori Cardellino's slide was the Ventura pier where I had met David. Both of our microphones were on as we discovered we both lived on the same street in Ventura, CA.

As we talked on the plane home, I told her all about our manuscript for *Plague of Corruption* and our first book *Plague*. Upon returning home she read *Plague*, and we began meeting for coffee at Kay's. After reading the manuscript for *Plague of Corruption*, she said simply, "You cannot publish that book that way."

A little put off as she talked to me in the exact tone and words of my late mother, I said "Why not?" somewhat defensively.

"It has no dead man's trigger," Dr Lori said emphatically.

I had no idea what that was so once she explained, I called up Kent and said, "Maybe there's one more story I should tell you." She next called her longtime friend Mikki Willis in Ojai and told him he must get my story on film.

Ever since Frank met Dr. Lori, he says, "Here comes Gloria," when he sees her. Dr Lori speaks to me with words and in a tone like my mom, Gloria. Most of the time it's funny.

Lori once told me of a near death experience where she was sent back to earth for me. As I often say, God has a sense of humor.

Dutifully, after Lori introduced us, Mikki Willis listened to me as I told my story. But this was almost a full year before anyone would know about COVID-19 and Mikki politely told me he was busy with other projects. I was relieved, as I certainly did not want to do a movie!

Another issue I was confronting during this time was the delay in publication of *Plague of Corruption*, from November 5, 2019 to April 14, 2020, in the midst of the COVID-19 plandemic. I don't pretend to understand the publishing world, and although Kent and I were very disappointed at the time, it seemed like God had His own plans. Readers eagerly gobbled up our tales of scientific misconduct and understood the ways in which our public health officials were lying to them about COVID-19.

Unwittingly, we'd given them the blueprint for the campaign of deception that Fauci and company were perpetuating on the world.

I'd sent a copy of *Plague of Corruption* to Joost Oppenheim at the NCI, as he knew better than anyone the persecution Frank had suffered at the hands of Robert Gallo, and his diminutive partner in crime, Anthony Fauci. I received a strange email back, in which Jo signed his name, "Joe," something I'd never seen in nearly four decades of my association with him. I was hoping that Jo would encourage Frank to read *Plague of Corruption* (which he eventually did), but prior to publication Frank would always beg off, saying "I don't need to read it. I lived it." The quick email response from "Joe" convinced me somebody other than Joost Oppenheim had intercepted my email. I was literally sick to my stomach when I read the response, in which he essentially wished me luck, but said he didn't agree with any of my assertions, and that the vaccine program was essentially beyond question. I don't believe Joost Oppenheim wrote that email.

Early in the plandemic, I was asked by the Academy of Nutritional Medicine to give a talk in early April 2020. The title of my talk was "Viruses Cause Disease by Dysregulating Key Immune Molecules Modulated by the eCS, The Dimmer Switch of Inflammation: Putting out the Fire."

As Frank helped me prepare for the talk, he emailed me a paper from January 2020 entitled "TGF-Beta Induced Dysregulation of SOCS3 Facilitates STAT3 Signaling to Promote Fibrosis."[1] Along with the article he sent a note, jokingly saying that our two favorite paradigm shifting discoveries were likely to be at the heart of solving this and perhaps every other pandemic associated with these families of deadly RNA viruses.

I put abstract right next to abstract right next to abstract of my work in the 1990s showing the difference between the pathogenic and non-pathogenic strains of Ebola. The cytokine storm profile from SARS-CoV-2 was the same as the pathogenic strain of Ebola. It was also a match for the cytokine storm of XMRV and ME/CFS (chronic fatigue syndrome).

I presented the talk online on April 5, 2020, because the entire country was starting our "fifteen days to flatten the curve" of COVID-19, which would last more than a year. Sadly, even the good people at the Academy of Nutritional Medicine told me I had to delete my slides on the influenza vaccines and inappropriate testing or else I wouldn't be allowed to speak.

I understood this pattern of disease, but nobody was letting me speak. I needed an angel to give me a voice.

God sent me several.

* * *

As Mikki recalls the meeting, "As I was listening to Judy's story, it became less relevant as to whether it was a narrative that would make a great movie. I became more focused on paying attention to her body language. The brokenness in her voice, what her eyes did when she talked about a painful situation, what her lips did. And I was clear by the end of her story, that she was telling the truth."[2] That short twenty-six minute video which was supposed to be a promotion of our *Plague of Corruption*, turned out to be one of the most watched videos of all time. A billion people saw it. Quickly Mikki and I both got death threats and were attacked by fact checkers so Mikki quickly began making the full-length movie. Mikki tells his story in a book to be published later in the Fall of 2021 called *Plandemic: The Incredible True Story about the Most Banned Documentary in History*.

While many expected the full *Plandemic* movie to feature me, *Plandemic* sent another angel, Dr. David Martin. I remembered seeing the opening minutes—me walking with Mikki while telling my story—and then the documentary pivoted to the Rotunda at the University of Virginia. I was confused as Mikki never filmed me at UVA. Enter Dr. David Martin, a brilliant man and innovator who'd developed technologies to track and detect white collar crime. Dr. Martin's take on the COVID-19 crisis was that it was best understood as a planned event, and he detailed the suspicious pattern of grants and patents which seemed to underlie the "plandemic."

The completed film, *Plandemic: Indoctornation*, received more than a hundred million views, and several million watched its debut on the London Real website, which seeks to be an outlet for traditional investigative journalism. Despite the vicious attacks on my video and the completed film, none of the critics ever stepped up to answer Mikki's ten-thousand-dollar challenge to disprove any of the claims in either video.

* * *

Documentary filmmaker Michael had both his Sicilian father and mother, independently of each other, send him the twenty-six-minute "Plandemic" video by Mikki Willis, and tell him he "had to watch it."

This was ironic, given his parents had never expressed a similar level of interest in his own films.

In an interview with my coauthor, Kent, Michael said, "I'm attracted to stories that are exposes of fundamental, systematic corruption and deception. I wish I was doing more human-driven stories. But I feel an obligation to tackle these big issues that cause a lot of cognitive dissonance for people."[3]

Michael went on to say:

> There's this big attitude that conspiracy theories are just for crazy people. But that's the equivalent of saying you don't believe in the mafia. Because another synonym for conspiracy theory is organized crime. So, if you don't like the term 'conspiracy theory,' just call it 'organized crime.' And we know organized crime is real and that's what my films deal with. And that's what you're dealing with in Judy's story. She's just one example of how the pharmaceutical mafia operates and destroys the life of anybody who comes into conflict with it.[4]

After watching the twenty-six minute "Plandemic" video, Michael said, "It resonated with me. It was sort of outrageous. But when I saw the media walking in lockstep to discredit it, and the social media companies working in lockstep to ban it, I immediately said, 'Okay, the establishment is terrified of this.' And whenever you see that you have to say to yourself, well, maybe there's something to this."[5]

He got a copy of our book, *Plague of Corruption*, and read it. He found the foreword by Robert F. Kennedy Jr. to be very compelling, and considered the main story of me and XMRV to be credible.

> Whether or not there was a cover-up with XMRV, a cover-up was certainly possible, because of the environment in which Judy was operating, which was unbelievably corrupt. And you can't really confront Judy's story, unless you understand that science is not the last guardian of objective knowledge. It's just as rotten and corrupt as the government, the media, and the intelligence agencies. Peer review is a disaster. My first start-up, a few years ago, was addressing the problems with peer review. So, this was something with which I was very familiar and a story I always wanted to tell.[6]

Michael and his team signed the deal with Skyhorse to make a documentary of *Plague of Corruption*. To answer the question of whether I was telling the truth, they hired one of the top investigative companies in the country, who often perform deep forensics for government agencies, celebrities, large companies, and billionaires. It was rumored that many of their senior members had worked for the National Security Agency (NSA). They scrutinized me, the claims made in our book, and the "Plandemic" video to determine if there was anything questionable.

Nothing came up.

But of course, Michael and his team wanted to be thorough, and what investigation isn't complete without a lie detector test, right? Michael gave me a call one day when I was in Carlsbad, CA and said, "Hey, Judy, can we give you a lie detector test in LA in about two hours?"

On a good day to drive from Carlsbad to LA in two hours was tough. But on a Friday afternoon? I whined a bit about traffic, but hey I'd only seen lie detector tests on TV, so I thought it sounded like fun. It didn't sound like fun to David but I promised him lunch at one of his favorite restaurants, Kristi's in Malibu, and he was game.

Michael told me he expected me to pass and was surprised when I accepted. As he later explained, "It was very last minute. And she said, 'Sure, I'll do it.' Absolutely no hesitation. We gave Judy a very rigorous lie detector test. It was four redundant lie detector tests in one. And she passed with flying colors. It really ruled out her being deceptive."[7]

However, trying to put together a well-balanced film, letting our side make our best case and the other side as well, didn't sit well with some people.

Maybe they were more used to our side not being able to speak. Michael said:

> We really tried to present the opposing view as strongly as we could. We've had some complaints from people involved in the production and distribution that the film is one-sided. And I said, 'Maybe it's because Judy, Frank, Robert Kennedy, Jr., and Kent are winning the argument. And that's making you uncomfortable.' Now, I didn't say they were winning the argument. But that's the conclusion a lot of people have drawn, and it makes them very uncomfortable. As the filmmakers, we're going to be neutral on Judy and Kent's story. You make up your own mind. But there was a lot of fishy stuff going on. And we know these pharmaceutical companies, and these regulators, they're criminal enterprises. So, it wouldn't surprise us if Judy and Kent were correct.[8]

The willingness to take on some very sacred cows led to some humorous moments in the shoot. The production had hired a music composer who was comfortable with the attack on Big Pharma in general but bristled when the film veered into criticism of Dr. Anthony Fauci. The composer said, "Fauci is worshipped like a god. Brad Pitt plays him on *Saturday Night Live*. You can't do that. This is a right-wing film."[9]

But an artist searches for truth, and this was no less the case for Michael. As he told Kent, "If through our investigations we had uncovered that she was full of shit, we would have buried her."[10]

Prior to interviewing John Coffin and Ian Lipkin, Michael showed his list of questions to Kent and asked if there were additional ones he should ask.

For each list, Kent added an additional eight to ten questions. After a decade of working with me, he knew where the scientific bodies had been buried by the establishment.

First up in the witness chair was John Coffin. A good interviewer makes his subject comfortable and Michael did an excellent job with his first few softball questions. As the interview continued and Kent's questions started being asked, Coffin said, "You guys have really done your homework on me."

As Michael described it, "He started being comfortable with us and really letting the mask slip and saying she had a terrible reputation. At the beginning of the interview, I asked if he was the peer reviewer on her *Science* paper. He said he shouldn't answer that, so he wouldn't. Later in the interview he said 'When I reviewed that paper, I didn't know Judy's reputation and she had a terrible reputation. She is known as the worst person to work with.'[11] He went even further later, saying I threatened to take a gun and shoot one of my critics.

When Coffin believed the filmmakers were on his side, that's when Michael said he "really started letting the mask slip."[12] In the beginning of the interview, Michael had questioned Coffin about his statement comparing me to Joan of Arc, when he claimed scientists would "burn her at the stake." He claimed not to remember making the comment, even though it was reported in *Science* magazine.

But when he believed Michael was on his side, a different, far nastier person appeared.

After the interview with Coffin, Michael called Kent to talk about whether that footage should be used. Kent told Michael, "The argument we've been making is that these scientists present one face to the public, but another in private. You now have evidence of that to show the world. I know it makes you uncomfortable, and you may get criticized for it. But people need to see this with their own eyes. The public's right to know about these people outweighs the embarrassment it might cause them."

* * *

By contrast, Ian Lipkin was much smoother than John Coffin.

Lipkin started off warm and friendly, and when my name came up said it was wrong I didn't have my notebooks but didn't know what he could do about it. He seemed to get a little worked up as Michael moved into Kent's questions, specifically on the question of whether XMRV would only intermittently be in the blood of the sickest patients, thus necessitating multiple blood draws over several months. "We didn't test for that," he grudgingly admitted.

I wasn't necessarily surprised by Lipkin's honest answer, as he often seems to shade the truth and confuse, rather than outright lie. As Michael's questions continued in difficulty, though, Lipkin became agitated. "Where are these questions coming from?" he asked at one point, then at another, "All these questions were answered years ago. We don't need to go over them again."

Michael did get Lipkin to admit on camera that the edict demanding that I not set foot on any NIH property under threat of arrest during the XMRV investigation came directly from Dr. Anthony Fauci. (We had emails from Lipkin detailing this arrangement.)

Michael also asked Lipkin about the Montoya samples he tested, in which he said on a CDC conference call on September 10, 2013:

> We found retroviruses in 85 percent of the sample pools. Again, it is very difficult at this point to know whether or not this is clinically significant. And given the previous experience with retroviruses in chronic fatigue, I am going to be very clear in telling you, although I am reporting this as present in Professor Montoya's samples, neither he nor we have concluded that there is a relationship to disease.[13]

Lipkin claimed he didn't remember saying such a thing on a CDC conference call.

And although Lipkin had told Michael he had all the time in the world, after a few more questions, he cut the interview short, saying, "I have a class I have to go teach."

The world's most celebrated virus hunter was driven off camera by the questions of my coauthor, an attorney and middle school science teacher.

* * *

As Michael eventually concluded, "What matters are her scientific claims. What's interesting is the science."

As Michael put it, "The question is whether our mechanism for the discovery of knowledge gets inherently hijacked by ideologues and corruption. And I think it does. Again, we're presenting the whole film as a conversation that's bigger than Judy. It's bigger than XMRV. XMRV is the jumping off point for us to look at this broken system."

Just as interesting as who would go on camera is those who would not. Michael emailed Annette Whittemore and received the following response:

From: Annette Whittemore
Date: September 25, 2020 at 12:07:15 PM PDT
To: Michael Binda
Subject: Re: CFS interview request

Hi Michael,
I'm happy to hear that you are doing a documentary about ME/CFS. The world needs to know what many scientists have learned over the years about the seriousness of this disease. The public needs to understand how much

the patients are still suffering from the lack of effective medical intervention. The Institute continues to serve this patient population, through its support of medical care, patient education, and outreach, in addition to its advocacy efforts to increase funding for research and clinical care. We are aided in these efforts by experts such as WPI Medical Director, Dr. Kenny DeMeirleir, and Associate Professor of Research at the University of Nevada, Reno, Dr. Vincent Lombardi, and by working together with many other nationally-recognized ME/CFS organizations.

The Institute has no comment on XMRV. However, you might be interested in interviewing Dr. Nath of the NIH on his efforts to discover the relationship between endogenous retroviral activation and chronic disease. He has advanced the study of this area of science in MS and ALS and is the lead NIH researcher in ME/CFS. His research focuses on understanding the pathophysiology of retroviral infections of the nervous system and the development of new diagnostic and therapeutic approaches for these diseases. Avi Nath, NINDS clinical director, is also developing a clinical protocol to study postinfectious myalgic encephalomyelopathy--chronic fatigue syndrome (ME-CFS).

(Look up endogenous retroviruses and you will understand why this would be a good place to look for evidence of disease.)

Good luck with your documentary!

Best Regards,
Annette Whittemore

How interesting that NIAID continues to fund the Institute and investigators who committed federal crimes of misappropriation of federal funds and admitted retroviruses play a role in ME/CFS, MS, and ALS, yet the meager funding for the disease comes not from the NIAID but from the National Institute of Neurodevelopmental Disease and Stroke (NINDS). They are not neurodevelopmental diseases if you inject the retroviruses from animals and aborted fetuses in contaminated vaccines!

Although it took Michael a long time to convince Frank to appear in the film (and though Frank agreed only to appear in shadows, his thick Boston Italian accent gives him away immediately and he also says his name repeatedly), it solidified in Michael's mind the two competing narratives about me.

There was "Saint Judy," who was right about everything, and "Crazy Judy," who was wrong about everything. And with Frank, he realized there's a nuance about me, just as there is with every human being. For example, regarding XMRV I said we absolutely discovered a new virus, and they covered it up. When Michael questioned Frank, his response was, "We may have discovered a new virus, but the replication studies were rigged to fail, so we don't know for sure."

As Michael said, "I asked her on camera, you say it this way, and Frank says it that way. Is he being too cautious, or are you not being cautious enough? She said without hesitation, 'Frank's right. I'm not being cautious enough.'"[14]

That's right. Perhaps I wasn't being cautious enough. But when even Frank says the replication studies were "rigged to fail," doesn't it mean they're terrified we were right? And if they were "rigged to fail," didn't that mean there was something they wanted to cover up?

Frank has long told people I'm the most "intuitive" scientist he's known. I take that as a compliment, and after I have my insight, I know the evidence needs to be assembled. I'm like the architect who sees the bridge before it's built, and Frank is the engineer who gets the job done. So, maybe I'll be a little more cautious in my words, for the benefit of those who don't yet see the bridge. But I still think I'm right. The data still support my interpretations and by the way my enemies act, it seems like they do as well.

In Michael's opinion, "At the end of the day, I find Judy and Frank's interpretation of what happened around XMRV to be persuasive. And what the film ends with is: we need an honest replication study. Because there's enough smoke here, that, well, maybe there's a fire."[15]

Spoken cautiously, like a true scientist.

But for this scientist, when I see the millions of people struggling with chronic conditions, it seems like there's a raging inferno out there.

CHAPTER FIVE

2008 Nobel Prize Winner Luc Montagnier—From World War II to COVID-19

Nothing in life is to be feared.
It is only to be understood.

—Marie Curie

In 1940, war came to France, and seven-year-old Luc Montagnier, along with his mother and father, fled the advancing German army, along with many others during this French exodus. Their home had been near a railway station and they thought it safer to run, but this led them to be more exposed to the German bombing. They would have been safer at home. The war years were terrible, with no food reserves, and Luc's family often found themselves starving.

In 1944, French Resistance fighters had alerted the Allies of a German fuel train near the Montagnier home. During the successful Allied bombing of this enemy fuel train, Luc's home was destroyed. In addition, soldiers of the retreating German army came to the farm where Luc's family had taken refuge, stealing his parents' two bicycles, as they tried to pedal furiously back to Germany.

The end of the war brought mixed feelings to young Luc.

Peace brought with it the liberation of the concentration camps and the emaciated survivors, survivors of a horror he could not even begin to imagine, many of whom were returning to France. During his primary school years, the young Luc was pushed ahead two grades because he was so advanced. Luc retains a vivid memory of being close to his thirteenth birthday when he heard the news of the atomic bomb being dropped on the city of Hiroshima on August 6, 1945, ushering in the nuclear age in a fireball that immediately killed over a hundred thousand people.

Prior to that time Luc had tried to lose himself in reading, devouring the works of Balzac and Voltaire, as well as the science fiction novels of Jules Verne with their tales of adventures in the stars. His father, an accountant, was also a voracious reader, preferring to spend his free time reading popular science books on topics ranging from physics to organic chemistry. But Luc realized his beloved science, instead of enlightening humanity, could also be used to usher in a dark age of terror under the threat of nuclear annihilation.

After the bombing of their house, the town government put the family into a modern, empty house which had been used by the local Gestapo during the war. In the cellar, Luc set up a small chemistry lab, producing hydrogen gas, sweet-smelling aldehydes, and nitro compounds. The proud parents of the young scientist were understandably concerned about the possibility of explosions. More than once, Luc proved the concern of his parents to have been a reasonable one.

Luc's parents wanted him to study literature as a prelude to becoming a lawyer, but he was interested in science. He'd considered becoming a physicist, but chose biology, which was called "natural sciences" in France at the time. However, his parents were pleased when he enrolled in a preparatory program in "Medicine" at the University of Poitiers, imagining he would someday become a doctor. The young Montagnier pursued a two-track strategy, taking medical courses, but also science classes, so that he might become a researcher. He found the science offerings of the Sorbonne to be disappointing, but frequenting the library of the famed Pasteur Institute challenged him to advance his scientific thinking.

At the age of twenty-three, Montagnier became an assistant in Cellular Biology at the Curie Institute, a laboratory specializing in cancer research. Fortunately, the Curie Institute had close ties with the Pasteur Institute. His interests shifted radically during this time, from plant biology and animal cells to viruses. Especially compelling to the young student was foot and mouth disease virus, an RNA virus.

In 1953, James Watson and Francis Crick (along with an uncredited Rosalind Franklin) discovered the double-helix, the so-called twisting ladder of DNA which carries the four nucleic bases, adenine, cytosine, thymine, and guanine. In 1957, the outlines of modern genetics were starting to become clear. Genetic information could be carried by either DNA (deoxyribonucleic acid) or RNA (ribonucleic acid.) DNA was used to carry genetic information for cells, while RNA was used only by certain viruses.

Around 1958, Montagnier spent several months developing an original technique for allowing RNA to enter the cells to determine if this RNA was enough to allow the virus to replicate. Exciting research during these years showed that certain viruses could replicate using only RNA, prompting many to wonder whether RNA might have a similar double-helix shape to DNA, or be using a completely different configuration.

In 1960, the future Nobel Laureate was already becoming acquainted with petty scientific power struggles. Montagnier's medical school thesis, "Infectious RNA of Bovine Foot and Mouth Diseases," was at first refused by his thesis adviser, Professor Fasquelle, a small-minded man. Soon thereafter, a more intelligent and open-minded professor recognized the importance of the work and willingly endorsed the thesis. But throughout his career, Montagnier found there were many small-minded people who tried to stymie scientific progress in various ways.

In July 1961, Luc was sent on a three-year scholarship to the laboratory of Kingsley Sanders, near London, where he would continue to investigate the mechanics of replication by viral RNA. After more than two years of hard work, often including weekends, he was the first to discern the shape of replicating viral RNA, a double helix similar to DNA, but much more rigid. This important discovery was published in the journal *Nature* in August 1963.[1] It was an elegant demonstration of Mother Nature's ability to recycle and improve upon useful designs. Many have theorized that RNA was used to transmit genetic information for the first life on Earth, or by the first molecules with some of the characteristics of life. When DNA made its appearance on the scene, it replaced RNA in all organisms we know, although every organism would retain RNA, in the form of messenger RNA, as the primary way of assembling proteins within the body.

In effect, RNA would be tasked by the master, DNA, to do a lot of the chores necessary to keep an organism alive and healthy.

* * *

From London, England, Montagnier moved to the University of Glasgow in Scotland, where the virologist, Michael Stocker, as well as the visiting Renato Dulbecco, were demonstrating how regular cells were turned into cancer cells by exposure to the polyoma virus. Montagnier demonstrated that both DNA and RNA viruses had the potential to create cancer cells, most notably causing changes in the plasma membrane of the cell, as well as the carbohydrate layer surrounding it.

One of the great mysteries of the time was how RNA viruses, particularly the retroviruses, were changing their genetic information to function in a DNA organism. Howard Temin had suggested some sort of enzyme was allowing for the transformation of RNA into DNA. This was eventually identified in 1970 as reverse transcriptase. Credit for this discovery was shared by Howard Temin, Satoshi Mizutani, and David Baltimore. This discovery led many to believe reverse transcriptase would be a good tool for identifying RNA viruses in leukemia and cancer in humans.

In the mid-1970s, Montagnier's research was on finding human retroviruses by the presence of reverse transcriptase when an unexpected discovery made this work much easier. Frank Ruscetti, working with Dennis Morgan in the Gallo lab, had discovered a growth factor for lymphocytes in cell culture, which would eventually be called Interleukin 2 (IL-2). This made Montagnier's cell cultures more durable, and made detection of retroviruses easier.

The search for the first disease-causing human retroviruses, HTLV-1 (human T-cell leukemia virus), was won by Frank Ruscetti, working with Bernie Poiesz in the Gallo lab in 1979.

Only a few short years later, Luc Montagnier and his team changed the history of the world with their discovery of HIV. Montagnier has often acknowledged the importance of Frank's Interleukin 2 in his isolation of HIV. In collaborating with Paris physicians, Montagnier received a biopsy sample of a swollen cervical lymph node of a young gay man who had recently traveled to the United States. Lymphadenopathy was already known as a precursor of potential serious illness, including cancer. In this case, the lymphadenopathy was a precursor of AIDS.

As Montagnier recounted for the biographical section of his Nobel Prize speech in 2008:

> The lymph node biopsy arrived on January 3, 1983, a date which I remember well because it was also the first day of the virology course at the Institute Pasteur, which I had to introduce. I could only dissect the small hard piece at the end of the day. I disassociated the lymphocytes with a Dounce glass homogenizer and started their stimulation in culture with a bacterial mitogen, Protein A, known as the activator of B and T

lymphocytes, since I did not know which fraction of the lymphocytes could produce the putative virus.

Three days later, I added the T-cell growth factor I had obtained from a colleague working in the laboratory of Jean Dausset. The T-cells grew well. As previously established, it was decided with my associates, Francoise Barre-Sinoussi and Jean-Claude Cermann, to measure the reverse transcriptase activity in the culture medium every three days. On day 15, Francoise showed me a hint of positivity (incorporation of radioactive thymidine in polymeric DNA), which was confirmed the following week.

We had evidence of a retrovirus, but this was just the beginning of a series of questions:

Was it close to HTLV or not?

Was it a passenger virus or, on the contrary, the real cause of the disease?

In order to answer these basic questions, we had to characterize the virus biochemically and immunologically, and to do that, we needed to propagate it in sufficient amounts. Fortunately, the virus could be easily propagated on activated T lymphocytes from adult blood donors. No cytopathic effect was observed with this first isolate. But unlike HTLV infected cultures, no trans-formed immortalized lines could emerge from these cultures, which always died after 3-4 weeks, as do normal lymphocytes.

By contrast, subsequent isolates I made from cultures of lymphocytes of sick patients with AIDS were cytopathic for T lymphocytes, and we discovered later, could be cultivated in larger amount in tumor cell lines derived from leukemia.

Shortly after the virus isolation, my co-workers and I were able to show that it was not immunologically related to HTLV, and in electron microscopy, it was very different from HTLV viral particles. In fact, as soon as June 1983, I noticed the quasi-identity of our virus with the published electron microscopy pictures of the Visna virus in sheep, the infectious anemia virus in horses, and the bovine lymphotropic virus: it was a retrolentivirus, a sub-family of viruses causing long-lasting disease in animals without immune-deficiency.

This indicated clearly that we were dealing with a virus very different from HTLV, and my task was to organize a team of researchers to accumulate evidence that this new virus was indeed the cause of AIDS.[2]

The Nobel Prize was a deserved crowning achievement for Luc Montagnier, a scientist who had first isolated and characterized the virus in young men with the disease.

However, there had been some bumps in the road, and it mostly involved the Americans. Around 1973, Montagnier met Robert Gallo, when they shared a room at a science conference held at an overbooked hotel.[3] Montagnier gives an abbreviated account of his dispute with Gallo and its subsequent resolution in his book, *Virus*, published in the United States in 2000:

> The NIH [National Institutes of Health] had taken out patents for HTLV-3 which were accepted by U.S. authorities, while the patent for LAV [Montagnier's original name for the virus was Lymphadenopathy Associated Virus], although submitted much earlier (in December 1983), had not been granted. A long legal battle ensued between the American and French teams, not to come to an end until March 1987, when a historic agreement was signed by the directors of the NIH and the Pasteur Institute and ratified by Ronald Reagan and Jacques Chirac. The two patents would become the joint property of the two institutions, which would share the royalties.[4]

I have to add that this is probably the first and last time in history that credit for a scientific discovery had to be approved by the leaders of two different countries. But such was the national pride on the line for each country as to who had discovered HIV. In an interview for this book, Montagnier mentioned how our previous book, *Plague of Corruption*, had given him some further insight into Gallo's character and professional life, including his vindictiveness toward Frank Ruscetti, who was getting credit for Interleukin 2, as well as the discovery of HTLV-1.

As Montagnier recalled his own interactions with Gallo:

> In June of 1983, Gallo and I were friends. A heated debate ensued at a Paris restaurant. I told him that the virus I had isolated was close to the lentivirus of animals. Gallo was not convinced. He insisted that my isolated virus must be from the HTLV-1 family. Despite the efforts of our mutual friend, Guy de The, we could not agree.[5]

Montagnier is careful not to ascribe any malicious intent to Gallo. However, many other scientists over the years have repeatedly questioned Gallo's purported usurpation of the discoveries of other researchers as his own.

One of the most admirable traits of Montagnier is how he remains a true scientist, knowing for certain only what the data tells him, and yet remaining free to theorize where the truth might be found. A question we discussed at length in our previous book, *Plague of Corruption,* is how HIV jumped into humans. The reigning theory is HIV came from chimpanzees, but how did it jump into humans? Montagnier does not share this view, believing HIV has been in humans for a long time, before it was "discovered." But unlike many supposed scientists, Montagnier does not stop the conversation, even when it veers into dangerous territory. I was endlessly pondering John Coffin's statement of July 22, 2009, when Frank heard him say, "Do you mean all those sequences we saw in the 1980s were real?"

Two theories exist, the first being a natural transmission from either humans eating an infected chimpanzee (known in Africa as "bush meat") or being bitten by one, the so-called "cut hunter" theory. However, we argued the more likely explanation involved the sacrifice of more than five hundred chimpanzees in the late 1950s in the development of an African polio shot given to more than a million residents of the continent. This hypothesis was covered extensively in the voluminous book, *The River: A Journey to the Source of HIV and AIDS,* by Edward Hooper, and published in 1999. As Montagnier writes in his book on this question:

> The origin of HIV-1 is even more mysterious. One is tempted to look for an animal source, as with HIV-2, some primate infected with the virus but not sick. For a long time this approach produced no results whatsoever: none of the viruses isolated in monkeys tested in Africa were close to HIV-1. The situation changed in the early 1990s, when one virus and then a second were isolated in chimpanzees.[6]

It starts to become clear to me why our investigation into the fourth known human Retrovirus family, XMRV, was so threatening to the establishment.

The question as to the origin of AIDS almost became a crisis with the publication of Hooper's book, claiming it was released into the human population through the use of chimpanzee tissue in African polio experiments. However, the "bush meat" theory or the "cut hunter" scenario appeared equally plausible. And besides, the scientific establishment wouldn't be expected to point the finger of blame at themselves. The question and the controversy around the origin of HIV were mostly forgotten.

With XMRV, a mouse virus which had somehow jumped into human beings, no plausible explanation other than the use of mouse tissue in medical research for vaccines and other products was possible.

People had not been eating "mice burgers."

And unlike the thrilling action adventure scenes of chimpanzee and human conflict depicted by writers like David Quammen in his book *The Chimp and the River: How AIDS Emerged from an African Forest*, there had been no heroic battle to the death between a jungle hunter and a wild mouse.

And as far as variants of XMRV, well, the same thing happened in HIV. I will be the first to admit I don't know the answers. But when the very questions are banished from the public arena, what other conclusion can be drawn than at the very least, some are very terrified to find out the answers? Montagnier detailed the continuing mysteries of HIV:

> Today, nine subtypes of the major group HIV-1 plus the new subgroup HIV-O have been identified, but the list is far from complete. It is therefore possible the virus existed in an endemic state (that is, at a low level) on several continents before developing into an epidemic.
>
> If we grant that the virus had been present in humans for a long time, in Africa and probably elsewhere, the question then arises as to why the epidemic is so recent. Why, in just a few years, has the same virus gone from being a sporadic infection to an explosive epidemic?[7]

Montagnier doesn't claim to know the answer to the questions he's posing, any more than I do. And yet, one is warned against venturing into such dangerous waters. With age comes skepticism, and although you might not know the truth, you get a pretty good sense of when people are lying about claims in your area of research. It wasn't surprising when Montagnier and I came to similar conclusions about the origins of the SARS-CoV2 virus responsible for the COVID-19 crisis. This is from a French newspaper in April 2020:

> Luc Montagnier, who won the Nobel Prize in 2008 for his work on HIV . . . said, "We have arrived at the conclusion that this virus was created." He accused "molecular biologists" of having inserted DNA sequences from HIV into a coronavirus, "probably" as part of their work to find a vaccine against AIDS.
>
> He said it was not clear how the virus had been able to escape the laboratory, and condemned scientists for doing "the work of a sorcerer's apprentice."
>
> Professor Montagnier has said that he is not the first to suggest the connection, and added that "a group of renowned Indian researchers" had also tried to publish a study showing that the new coronavirus includes HIV DNA, but were forced to retract their claims and had been "smothered."[8]

As a molecular biologist, I appreciate that in order to study retroviruses one must be able to grow them. My entire career in the lab has been focused on developing technologies to produce large quantities of retroviruses for study.

Those technologies are immortalized cell lines.

I had worked in the 1990s with the Vero monkey kidney cell line, culturing Ebola viruses.

Viruses don't "escape" laboratories.

They are shipped around the world in those cell lines and walk out in the infected lab workers.

I think one of the signs of a truly great mind is understanding that even though you may have a deep understanding of your own subject matter, one is continually interested in learning new things. Even though Montagnier is known as one of the world's foremost retroviral hunters, he is unsatisfied with the lingering questions which remain, such as why cells die prior to the infection becoming severe, a condition known as apoptosis. At the end of his biographical section for the Nobel Prize, Montagnier states the challenge in the following way:

> We have spent a lot of time trying to find the origin of this massive apoptosis, without finding a completely satisfactory explanation: the most likely is the intensive oxidative stress existing in patients since the beginning of their infection. This is also a finding of which I am very proud: although oxidative stress has been – and still is – completely overlooked by AIDS researchers, it is likely to aggravate the wrong activation of the immune system at the origin of its decline and also it triggers inflammation through the production of cytokines.
>
> Of course, the next question arises: what are the factors causing oxidative stress; viral proteins, fragments of viral DNA, co-infection with mycoplasmas? Even after 25 years, we still do not know the complete answer. But the phenomena does exist and needs to be treated, while most AIDS clinicians do not care about it at all! The treatment by combined retroviral therapy has, without doubt, changed the prognosis of this lethal disease, from a death sentence to an almost "normal" life. However, the virus is still there, ready to multiply when the treatment is interrupted, and not all HIV infected patients in the developing world have access to it. And the epidemic still kills 2-3 million people a year. It is thus absolutely necessary to resolve these problems.[9]

A scientist stays on a problem until all of the questions are answered. There have been great strides made in HIV/AIDS. It isn't enough. Just as we have not answered critical questions about chronic fatigue syndrome (ME/CFS), autism, or even cancer.

We must do more to answer these questions.

* * *

When one enters a controversy, you discover your true friends and enemies.

Yes, it was possible to believe in those heady days of 2009 and through much of 2010 that the public health authorities would take the appropriate action on behalf of the chronic fatigue syndrome community which had been neglected for more than two decades.

The dangerous question I asked was, "If we are growing viruses in animal tissue, how do we know that other viruses in the animal tissue aren't coming back in the vaccine, or pharmaceutical product, which is being developed?"

However, while much of the scientific establishment and the mainstream media joined the witch hunt against me, Luc Montagnier has always been a stalwart champion of my right to ask questions and go wherever the data leads. He was brave enough to endorse our book, *Plague of Corruption*, and has consistently supported my research.

Montagnier is careful not to say he always agrees with my conclusions, but that the scientific process must be followed, and all objections discussed in an open manner, without the public shaming of me that has taken place. However, he acknowledges the risks I have pointed out. This is from my coauthor's interview with Dr. Montagnier:

> **Kent:** Is it accurate to say you don't have much of an opinion, either positive or negative, about Judy's XMRV work?
>
> **Dr. Montagnier:** Yes. But I have to say there is no reason to doubt Judy Mikovits. She had problems because she mentioned vaccines. I have my own concerns about vaccines. She was concerned that some viral material may have been contained in vaccines. I have been concerned about bacteria from plants contaminating the vaccines. They have been found in human cells.
>
> **Kent:** Do you consider it a theoretical possibility that since we grow human viruses in animal tissues, that when we create a product from this mixing of human and animal tissue, that we may have unsuspected animal viruses coming back from that product?

Dr. Montagnier: Yes, it is quite possible. We are so paralyzed.

Kent: Should science be actively investigating whether there are animal viruses coming back in these products which are developed for use by humans?

Dr. Montagnier: Quite assuredly, yes. Especially if we use live attenuated vaccines, like in the case of MMR (measles-mumps-rubella.) It is quite possible there could be transmission. We could also have contamination by bacteria. We are changing our environment.[10]

In a single exchange with the Nobel Prize winner responsible for identifying our greatest modern plague, HIV/AIDS, he tells you that what has happened to me has been wrong.

However, like me, Luc Montagnier is an optimist: "I think we are at a turning point which may revolutionize medicine. The era now is made of chemicals. We might have things in the vaccines. I think there is no solution but changing the chemicals that are causing these problems and chronic diseases. We have proved that in many cases, these chemicals are causing problems."[11] While my expertise is molecular biology, I cannot help but agree with him that many of the chemicals we are currently using are affecting our health.

And like me, Dr. Montagnier doesn't worry that his comments might cause him to be "canceled" in our current media environment. A scientist's obligation is only to the search for truth. That is why free speech is so essential:

I am quite convinced that we should not do what we're doing now. We should turn other ways. Let's take the vaccine story now. The new RNA vaccines for COVID-19 are quite a new innovation. Maybe it's too early. Maybe they're going too fast. It might have some applications in the future. But we have to be very careful because if we touch our genome, we have to know more about this. We know only 3 percent of our genes are coding. The other 97 percent, we don't know what it's doing. Are we going to change our organs with these vaccines? It's a new thing. Nobody can predict what is going to happen. Maybe more people are going to die of infectious diseases.[12]

A true scientist is humble before God's nature. There's no room for arrogance or hubris. When reasonable questions are excluded from the public debate we're not looking at science. We're in the realm of politics, persuasion, and big money.

There are still many true scientists/scholars among us. Several of those, who have investigated some of those chemical and biological toxins contributing to the explosion of cancer, autoimmune, and neuro-immune diseases, contacted me after seeing "Plandemic" or reading our books, *Plague* and *Plague of Corruption*. They graciously offered to lend their expertise to solutions, regardless of whether we might disagree on the cause.

We are all grateful to have the support of Dr. Luc Montagnier.

CHAPTER SIX

Environmental Toxins and Oxidative Stress Fuel Retroviral Associated Diseases

A wise man will make more opportunities than he finds.

—Francis Bacon

There have been scientists and physicians who honored the obligation to produce knowledge and communicate it despite the cost in this age of corruption.

We've discussed in previous books people like Dr. Jeff Bradstreet and Dr. Timothy Cunningham. Many more brave individuals contacted me after seeing the "Plandemic" video or reading our book. That is one of the great blessings of the past year.

For me, professionally, 2020 was my best year in more than a decade.

Scholars called and emailed me wanting to talk about science and solutions for COVID and many other conditions, seeing many of the same connections that Luc Montagnier discussed with my coauthor, Kent.

* * *

For most mortals, a single degree from the Massachusetts Institute of Technology (MIT) would be enough of an accomplishment to brag about for a lifetime. However, Dr. Stephanie Seneff has four degrees from MIT, in addition to being a senior research scientist at the institute. On the research side this is equivalent to be being a full, tenured professor. From the brief biography of her curriculum vitae:

> Stephanie Seneff is a Senior Research Scientist at MIT's Computer Science and Artificial Intelligence Laboratory. She has a Bachelor's degree in biology with a minor in food and nutrition, and a Master's degree, an Engineer's degree, and a PhD in Electrical Engineering and Computer Science, all from MIT. Throughout her career, Dr. Seneff has conducted research in diverse areas, including human auditory modeling, spoken dialogue systems, natural language processing, information retrieval and summarization, and computational biology, among others. She has published over 200 refereed articles in technical journals and conference proceedings on these subjects and has been invited to give many keynote speeches.
>
> Since 2008, Dr. Seneff has become interested in the effect of drugs, toxic chemicals and diet on human health and disease, and she has written and spoken extensively, articulating her view on these subjects. In particular, she has authored over 35 recently published peer-reviewed papers on theories proposing that a low micronutrient, high-carbohydrate diet contributes to the metabolic syndrome and to Alzheimer's disease, and the sulfur deficiency, environmental toxicants, and insufficient sunlight exposure to the skin and eyes play an important role in many modern conditions and diseases, including heart disease, diabetes, gastrointestinal problems, Alzheimer's disease, and autism. She has zeroed in on the herbicide glyphosate as being a major contributor to the alarming rise we are witnessing recently in multiple auto-immune, oncological, metabolic, and neurological diseases.[1]

By any objective measure the career of Dr. Stephanie Seneff has been an intellectual adventure and she has embraced that journey without fear, regardless of where it has taken her. She first stepped foot on the MIT campus as an undergraduate in 1964, and is still working there, even though she says with a laugh, many people claim she's retired. In her opinion, she's working harder than ever.

As she pursued her career, she found her greatest strength was the ability to understand large volumes of data, becoming in many ways the classic "data geek," seeing patterns in the information long before they became visible to others. Unlike many academics who studied health, the fact that her computer science research was not supported in any way by the pharmaceutical industry gave her an enormous amount of freedom to pursue this question.

One of the first associations to pop out from the data was a correlation between the MMR (measles-mumps-rubella) vaccine and autism. That puzzled her as the MMR shot used a live, but weakened virus, and therefore did not contain any mercury or aluminum, two of the most common suspected culprits behind autism. Seneff was so sheltered from the autism storms that at the time she discovered this link she was completely unaware of Dr. Andrew Wakefield, but recalled she thought MMR is a critical factor in autism. As she later recalled:

> However, after five years, learning about all the complicated and interesting comorbidities around autism, I still didn't feel I had the answer. I knew vaccines were connected. But I didn't think it was the main story. I went looking for something else. That was when it was serendipitous that I just happened to be at a conference with Dr. Don Huber, who is right at the top of my personal list of science heroes. [I only heard Dr. Huber speak once, in 2019, and he is also one of my personal heroes.] He was going to be giving a talk on glyphosate. I didn't know what glyphosate was, but I thought it looked interesting. It was a two-hour presentation on glyphosate and I walked away from that talk convinced I'd found the answer. It was that clear.
>
> I didn't know what glyphosate was, but I knew it was the main ingredient in Roundup. I quickly found out it's the most used herbicide on the planet. And the United States uses more of it than anybody else in the world. Huber talked about glyphosate causing gut problems. He talked about the soil bacteria being linked to the gut bacteria and disrupting the gut microbiome. And I knew the gut microbiome was linked to autism. And he talked about chelating minerals and autism kids have difficulty with manganese, iron, and zinc.
>
> Glyphosate disrupts the body's ability to arrange the minerals properly. Iron and manganese can both be toxic. But they're also essential. So, if something is toxic and essential, glyphosate must be putting a monkey wrench into the whole system. You end up with toxicity and deficiency at the same time because you can't get the numbers right. And you can't just take more to fix the problem.[2]

We met at the Autism One Conference in 2010, the same meeting where I met Kent Heckenlively. Autism One was one of the three organizations that never stopped inviting me to participate in annual conferences, regardless of any of the negative stories told about me. Stephanie's slides were as complex as mine, but Stephanie had the gift of explaining the Shikimate pathway much better than I have ever explained DNA methylation. This was a woman who thought about a problem and didn't simply settle for the first thing that made sense.

Appreciating the complexity of chronic diseases, Seneff knew it was critical to ask the right questions. Questions like: what are the mechanisms? And, are there any other contributing factors? However, scientists in the health field are "paralyzed" because asking these questions puts pharma dollars at enormous risk.

Roundup is the most commonly used herbicide (and is also sprayed on crops as a drying agent) and it does a really good job of killing things, including the beneficial microbes you're supposed to consume along with the produce. So, what happens to the community of microbes in your gut, better known as the "microbiome," which is supposed to produce a lot of the nutrients necessary for your good heath?

The answer is glyphosate decimates your microbiome.

If you don't have the proper type of microbes found in a typical, healthy microbiome, you won't be able to generate the proper level of things like iron from your food and will suffer from anemia. Kent and I have spent a decade writing these books to bring attention to the three-decades-long plague of corruption perpetrated by the FDA, CDC, and Fauci, similar to what happened with HIV/AIDS. Glyphosate is not the most toxic chemical around, but because people have been told that it's almost as safe as table salt, they use it so casually, spraying it on dandelions while kids are playing in the yard.

We've become aware that gut microbes are really important for our health in many diseases, including autism, Alzheimer's diseases, inflammatory bowel disease, liver problems, arthritis, and Parkinson's disease. They've been shown to trace back to the gut and the microbiome community, which communicates with the brain. It controls a whole bunch of parameters that influence activity of the metabolic systems. There's a lot of crosstalk between the gut and the brain.

The gut microbiome also produces a lot of essential nutrients for the host. And many of those essential nutrients that are produced come out of those specific pathways that glyphosate disrupts. It's been pretty well-established that the enzyme that gets hit hardest by glyphosate, the thing that kills the plant, is an enzyme called EPSP synthase.

And that enzyme is in a biological pathway called the shikimate pathway. And the shikimate pathway is present in pretty much all plants and many microbes. The pathway is used to produce the aromatic amino acids, tryptophan, tyrosine, and phenylalanine. The aromatic amino acids are coding amino acids, which means they go into the proteins. So, if you have a deficiency in those amino acids, then you've got a problem with protein synthesis, and that can have a lot of consequences.[3]

Consider the importance of what Seneff is saying. A common thread between autism, HIV/AIDS, cancer, autoimmune disease, ME/CFS, and Lyme disease is a lack of diversity of microbes in the gut microbiome. In other words, diversity in the gut is a sign of health, and a lack of diversity is a precursor to disease. Simply stated, we need those microbes in the gut that glyphosate is killing to prevent retrovirus-associated diseases like AIDS and ME/CFS.

Seneff and I both appreciate correlation is not causation. Once you see a correlation, then you go looking for a mechanism which might explain it. And we have that explanation. We know that glyphosate kills plants by interfering with the enzyme, EPSP synthase, and messing up the Shikimate pathway. Why should we believe when we ingest glyphosate it would be any different with us? Seneff continued:

> These aromatic amino acids are also precursors to neurotransmitters, like dopamine, serotonin, and melatonin. [These are pathways also dysregulated in human retrovirus associated diseases.] They're also precursors to the skin-tanning agent melanin and the thyroid hormones, and some B vitamins. If your microbes are affected, such that there are troubles in the shikimate pathway, you're going to have deficiencies in the production of all these compounds, and you have deficiencies in these compounds in all these diseases.[4]

Seneff and I appreciated the possible synergies in retroviral dysregulation of DNA methylation pathways and glyphosate dysregulation of Shikimate pathways.

I learned so much about the mechanisms of glyphosate toxicity from the Seneff lectures I couldn't write fast enough. Seneff had many important slides, but when I heard her mention methionine and hydrogen sulfide gas, I really perked up.

Seneff goes on to detail other ways that this bacterial overgrowth of the wrong species can cause problems. Glyphosate also disrupts the synthesis of methionine, a sulfur-containing amino acid, by the gut microbes. Instead, they produce hydrogen sulfide gas, and this causes other problems like bloating and abdominal pain. Hydrogen sulfide gas is also often generated as a result, which leads to methionine deficiency, another common problem in children with autism. Seneff explains:

> Methionine is a sulfur containing amino acid and it's essential. And it's a precursor to glutathione. And when you get a glutathione deficiency because of the methionine deficiency you get liver damage, because of the oxidative damage. [Methionine deficiency and hydrogen sulfide gas production is a common issue in the ME/CFS patients with XMRV and the inflammatory cytokine signature of XMRVs. Of course, I was thinking of the methyl donor, S-adenosylmethionine.]
>
> You get this whole cascade of health problems. In this recent study on rats who were exposed to glyphosate levels below the regulatory limit, they got fatty liver disease. Glyphosate disrupts the liver metabolism because it actually blocks the bile acid production and disrupts the cytochrome P-450 enzymes, which are crucial for activating vitamin D, which is also linked to autism.[5]

The Autism One conferences have always been cutting edge science and applied treatment strategies. It's a tragedy to me that such well-reasoned explanations never make their way into mainstream medicine or media. Glutathione is known as a detoxifier of heavy metals and pesticides. But with glyphosate what happens is those compounds simply stay in the body.

I understand that most doctors don't have the time to do much more than glance at the day's headlines or journal abstracts. They're expecting that scientists and public health officials are doing their job with integrity. Sadly, this is not true.

To get the word out on glyphosate, Dr. Seneff is writing her own book, which came out in mid-2021, called *Toxic Legacy: How the Weedkiller Glyphosate Is Destroying Our Health and the Environment*. In the book, she adds an additional concern about glyphosate. That glyphosate can fool the DNA code, pretending to be the amino acid glycine. This causes it to get inserted into proteins in a spot where glycine was intended. Glycine is the smallest amino acid, and it plays an essential role in many proteins.

> Glyphosate is an amino acid, and that's important. Most toxic chemicals are not amino acids. It's an unusual aspect of glyphosate. It's also an amino acid

analog of glycine, and glycine is one of twenty or so coding amino acids. The DNA code uses four letters, represented by four nucleotides, which code in sequential three letter units for the assembly of proteins, like beads on a string. Glycine has many essential roles in various proteins. So, when glyphosate gets substituted for glycine, you're changing the protein. In my book, I show what I call the glyphosate susceptibility motif. It's a very specific context in which I think glyphosate is very, very dangerous. And if I'm right, I can predict which proteins would be most affected by glyphosate.[6]

In a 2018 article in collaboration with a researcher from Boston University, Seneff laid out this theory:

> One possibility is that glyphosate, acting as an amino acid analogue of glycine, can erroneously become incorporated into proteins in place of glycine. This would lead to a cumulative and insidious toxic effect. Glyphosate is a glycine molecule with a methyl phosphonate group attached to its nitrogen atom. Proline is a coding amino acid that, like glyphosate, has an additional carbon bond on the nitrogen atom. But this does not preclude its incorporation into a peptide chain, demonstrating that glyphosate could do the same. Because its core structure is a glycine molecule, glyphosate can potentially be misinterpreted as glycine, based on an apparent match to the DNA code for glycine.[7]

I'm a chemist so to me this is a no-brainer. Glycine incorporates into the proteins which are the workhorses of the body.

Glyphosate is a modified glycine molecule.

In protein synthesis, it's easy for glyphosate to be mistaken for glycine. Therefore, instead of having your proteins built with glycine, a naturally occurring biological molecule, you're building them with an herbicide. Yet apparently, they don't have chemists at Monsanto.

And of course, this process is slow and insidious, not fast and acute. That's one of the problems with glyphosate because you don't immediately know you've been poisoned. It takes time. It's a slow kill. This reminds me of aspartame and Dr. Betty Martin's heroic efforts to get this slow, insidious killer poison out of our food supply. Yet again, even outlawed, corrupt Big Pharma and the FDA look the other way and deny. Slow, insidious poisons could absolutely synergize with slow, insidious lentiviruses or bacteria like Borellia to cause chronic diseases, as theorized by Luc Montagnier and myself. Mainstream medicine and media want to force the principle of Occam's Razor to deny the complexity of the causation behind these chronic diseases.

As Seneff further explains:

> That's how it can cause all sorts of things. The rise in glyphosate usage over time matches incredibly well with the rise in a long list of debilitating diseases. The correlation coefficients are over 0.95 and the p-value has several zeroes before you get to the first non-zero digit, (such as 0.00000073). So it is very tiny in these correlation calculations. These correlations are stunning.[8]

We had an even more stunning p-value in our original *Science* paper. I think we had more than ten zeroes in front of one of the associations of the new family of gamma retroviruses to ME/CFS. That means there was a one in ten billion chance that the virus was NOT associated with the diseases. We never said it proved causation, because we appreciate there were many potential factors involved.

Seneff sent me papers she'd written with Nancy Swanson, showing a list of diseases that are going up in concert with glyphosate usage. Alzheimer's, autism, Parkinson's disease, inflammatory bowel disease, liver disease, kidney disease, thyroid cancer, and pancreatic cancer, among others. All of these have gone up dramatically.

Of course, the critics will say, "Oh, everything correlates with everything else." They've just been amazing in their ability to divide and isolate us from each other so we don't see these synergies. I look forward to Seneff's book where she takes each and every one of these diseases and shows exactly how glyphosate can affect specific proteins.

My personal favorites are the DNA methyl transferases.

Glyphosate is one chemical that had not been on my radar for its ability to disrupt the proper functioning of the immune system.

Seneff believes the reason the United States has had such a difficult time with COVID-19 is due to our heavy usage of glyphosate.

SARS-CoV-2 is a single stranded RNA virus. There is a low level of expression of the virome all the time, and, in the person, those HERVs can convert RNA into DNA, and it gets integrated into your genome. We now know that 8 percent of our genome is a virome (viruses integrated into your genetic code) and the endogenous gamma retrovirus (HERVWenv), syncytin, is part of the SARS-CoV-2 spike. These sequences can by acquired into the genome and recombine into new variants, defective or infectious viruses, all contributing to disease development. As Seneff explains:

> That's true for the mRNA vaccine as well. The vaccine has the RNA. The RNA can then be converted to DNA by your retroviruses. Then the DNA can be integrated into the genome of your immune cells, your stem cells, and you can go along producing the protein forever.[9]

Therefore, the concern is that people with undiagnosed retroviruses, like XMRVs, are at an elevated risk if they come into contact with SARS-CoV-2. The viruses, in effect, help each other to become more dangerous.

Not only might these people be exposed to the RNA of SARS-CoV-2, but we're giving them RNA in the COVID-19 vaccine. Seneff continues:

> There was a recent article in *Nature*,[10] about a patient who had cancer and they'd given him a chemotherapy drug that suppressed the immune system. He was hospitalized with COVID-19, and subsequently he got serum antibodies from people who'd recovered. He stayed alive for a hundred and one days. But for that entire time, he was contagious. He couldn't fight the virus off and it kept mutating. When he died, his dominant strain was a variant that had twelve different mutations in the spike protein.
>
> They theorized this is what was happening with the variants that are now popping up around the world, like what's coming out of the United Kingdom. A person with a weakened immune system gets COVID-19, then they might get antibodies. But their innate immune system is so frail that the antibodies can't clear the virus. This sets up a situation that encourages the virus to mutate into a form that resists the antibodies.[11]

Seneff expressed the same concern to my coauthor Kent about the massive vaccination of the immune-compromised population.

An additional concern is that the very spike protein the vaccine is attempting to elicit antibodies into the body, is linked to many autoimmune conditions, especially if it persists in the body. In how many ways is the COVID-19 vaccine likely to be a problem?

Another likely problem with any coronavirus vaccine is the issue with antibody dependent enhancement. Let me spend a minute explaining what I mean, because it may be the strongest argument against any type of coronavirus vaccine. The reason you want to catch a virus, assuming you're in good health, is that your immune system will create antibodies to fight off that virus in the future, and also any likely variants.

Think of it like strength-training for your immune system.

However, something extremely odd seems to happen when researchers in the past have created a potential vaccine against a coronavirus. The animals get the vaccine, their immune system produces the antibodies expected to protect against the virus, but then when they get exposed to another coronavirus, their immune system violently overreacted, causing disease. This is an article about the problem, published in *PLOSOne*, in 2012, about the effort to develop a vaccine to deal with the SARS (Severe Acute Respiratory Syndrome) virus in China in 2002:

> For this purpose, the National Institute for Allergy and Infectious Diseases supported preparation of vaccines for evaluation for potential use in humans. This effort was hampered by the occurrence in the initial preclinical trial of an immuno-pathologic-type lung disease among ferrets and Cynomologus monkeys given a whole virus vaccine adjuvanted with alum and challenged with SARS-CoV. That lung disease exhibited the characteristics of a Th-2-type immune-pathology with eosinophils in the lung section suggesting hypersensitivity that was reminiscent of the descriptions of the Th-2-type immune-pathologic reaction in young children given an inactivated RSV vaccine and subsequently infected with naturally occurring RSV. Most of these children experienced severe disease with infection that led to a high frequency of hospitalizations; two children died from infection. The conclusion from that experience was clear; RSV lung disease was enhanced by prior vaccination.[12]

You should know that RSV stands for Respiratory Syncytial Virus, a single-stranded RNA virus that causes respiratory problems, and thus is similar in many ways to a coronavirus. Taking the lead on these vaccines was NIAID, run by Dr. Anthony Fauci, who has been leading our nation's efforts against the coronavirus crisis. Therefore, it's a pretty good bet he's as familiar with this research as Stephanie and I are.

In plain English, the research showed that vaccination ended up causing the very diseases that were trying to be prevented by the vaccine, when exposed to a similar virus. Those who got the vaccine ended up having a more severe reaction, especially along the Th-2 pathway discussed by Frank Shallenberger, which affected their lungs and breathing.

The innate immune system is messed up because of glyphosate and all the other chemicals in our environment. The virus comes in, and what does it do? It strengthens the innate immune system. But because the innate immune system can't fight it off, the virus multiplies like crazy. The immune cells swarm into your lungs, as well as platelets, and each one of the platelets has from five to eight mitochondria.

The platelets release their mitochondria and the immune cells take them up. The platelets are in essence saying to the immune cells, 'Here, take my mitochondria. You need them more than I do.' The immune cells are able to restore their health through this process, if you survive. If all goes well, you get really sick, then you recover, and your immune system is stronger. If your immune system was already strong, you won't get sick.[13]

Science is about confronting the unknown, wrestling with the questions, and coming up with solutions.

When asked about solutions, Seneff is clear in stating there should be a worldwide ban on Roundup, as well as the use of glyphosate in all products. It's much too dangerous a molecule with the ability to mess up our genetic code to be continued to exist in our environment.

Glyphosate should play no meaningful role for plants, insects, humans, or any life on Earth.

* * *

I also first learned of the work of Dr. Christopher Shaw, a Canadian neuro-scientist and professor at the University of British Columbia in Vancouver, Canada, at an Autism One conference in 2015. There, a parent had asked me to serve as an expert in her child's case in the National Vaccine Injury Compensation program. Drs. Shaw and Christopher Exley were also serving as experts on the case. It was the first time I realized how toxic injected aluminum was to the brain macrophage (microglia).

Dr. Shaw has served in the armies of two countries. His father and uncles served in World War II, and despite growing up in a non-religious Jewish household, there was still a strong secular moral sense of needing to do the right thing. From an early age, Chris was deeply moved by the Hebrew admonition, *tzedek, tzedek tirdof*, translated as "Justice! Justice, thou shalt pursue!" from Deuteronomy. The religious tenets of Judaism were not central to Chris's development as a young man, but the ethical teachings made a lasting impression.

He graduated from the University of California, Irvine in 1971 with a bachelor of science in neurobiology, and then went to Hebrew University in Jerusalem to get his master's degree in medical physiology. He remained at Hebrew University to get a doctorate in neurobiology and zoology, and did his post-graduate work with a vision research group at Dalhousie University in Nova Scotia, before being recruited to the University of British Columbia in 1988.

At the time Chris started at the University of British Columbia he was looking at how chemical neuro-transmitters were being received by cellular receptors, and whether this activity could be affected by various electrical and chemical interventions. Around this time, Chris started working on ALS (amyotrophic lateral sclerosis), also known as Lou Gehrig's disease. In ALS, the body itself seems to mount an attack on cells in the brain and spinal cord, leading to muscle weakness and paralysis. In the final stages, patients often develop respiratory failure and die from pneumonia.

Eventually, Chris and his team began to focus on a very strange cluster of ALS and Parkinson's disease (another disease that attacks nerve cells in the brain, resulting in a loss of muscle function, often resulting in tremors) in Guam. Parkinson's is commonly referred to as a non-fatal disease, although one often dies from "complications" related to the Parkinson's, generally lowering the expected lifespan of a person by one to two years. During the 1950s and 1960s, this strange ALS-Parkinson's like disease was one of the major causes of death on the island.

Chris and his team discovered strong evidence that the Guam cluster was linked to various types of neurotoxins in the food, specifically the use of immature cycad seeds in the production of flour, something that happened in the years after World War II. In June 2002, Shaw and his team published a detailed explanation of their hypothesis, backed up by an animal model.[14]

In the 2004 to 2005 time-frame, Chris and his team were looking for another cluster of ALS and ran across studies on veterans from the First Gulf War. The ALS cluster among the Gulf War veterans was first identified by Dr. Boyd Haley, the professor of chemistry from the University of Kentucky. The thinking was that the condition in these military veterans might be caused by either the anti-nerve gas agents they'd been given, or the large number of vaccines they'd been given prior to deployment, specifically the anthrax vaccine.

It had been suggested that the aluminum adjuvant in the vaccine might be the problem, so Shaw came up with a simple experiment. He would inject a size-proportionate dose of the aluminum adjuvant into the mice, and watch what happened. This is what Chris and his research team reported in their 2007 publication:

> Gulf War illness (GWI) affects a significant percentage of veterans of the 1991 conflict, but its origin remains unknown. Associated with some cases of GWI are increased incidences of amyotrophic lateral sclerosis and other neurological disorders. Whereas many environmental factors have been linked to GWI, the role of the anthrax vaccine has come under increasing scrutiny. Among the vaccine's potentially toxic components are the adjuvants aluminum hydroxide and squalene. To examine whether these compounds might contribute to neuronal deficits associated with GWI, an animal model for examining the potential neurological impact of aluminum hydroxide, squalene, or aluminum hydroxide combined with squalene was developed . . .
>
> The findings suggest a possible role for the aluminum adjuvant in some neurological features associated with GWI and possibly an additional role for the combination of adjuvants.[15]

As Chris testified in the hearing, the inevitable question became, if the aluminum hydroxide was being used in vaccines for children (about two-thirds of the vaccines for children have aluminum), what conditions might it be causing?

One of the arguments that Chris does an excellent job addressing is whether aluminum should be considered dangerous as it is the most common metal in the Earth's crust and the third most common element. He explains,

> During the industrial revolution we learned how to extract aluminum from bauxite and other minerals. Biologically, it was never something life on Earth had to deal with until the 1830s and 1840s. It has really changed our biosphere. If you eat it, and you have good kidney function, you'll mostly excrete it. It comes out through your urine, your feces, and your sweat. However, if you inject it, the trajectory in the body is very different. It gets picked up by macrophages (immune defense cells), hauled around the body, and deposited in one of two places, your bones or your brain. So, when people say eating it is the same thing as injecting it, well, no, it's not.[16]

Because of all the varying ideas to which Shaw was being exposed, he eventually decided to write a book about the various schools of thought on what was behind the rise in neurodevelopmental problems in our population, particularly among the young and the elderly. Shaw's book is titled *Dispatches from the Vaccine Wars*, and is scheduled to be published in 2021.

As Chris told Kent while interviewing for this book:

> Writing the book has really helped broaden my perspective on the whole thing and the role of Big Pharma. Especially what's called 'exceptions' and mandates and all those sorts of things. Reading your book with Judy made it clear how corrupt a lot of the science has become, particularly in vaccinology.
>
> And the corruption in medical education, the associated power of pharma, and what we call 'regulatory capture.' They are very powerful. They have basically taken this branch of science and made it a secular religion. The idea that you can question it is very dangerous in this field. Because people who do question it tend to get punished or get fired. Even people who are 90% pure, if they deviate, they get punished, too . . .
>
> That's where I see the relationship to religion, and it's not a positive comparison.[17]

Like Chris, my definition of science sees it as a process of inquiry, not a bunch of inflexible rules handed down from on high.

As Chris says of the current situation,

> You could compare it to going up against the Catholic Church five hundred years ago, or the Islamic State very recently. If you deviate from the true faith, you're going to get it. And I think that's driving a lot of the reluctance from various researchers to fully engage with this issue. The way it's evolved in the past couple of years, you'd have to be pretty crazy to go into any independent research on aluminum or vaccines.[18]

I guess I appreciate why the media continues to call me crazy, just as they did a decade ago. I guess it takes a special kind of crazy not to give up in the face of all the attacks which have been leveled against me since 2011.

It's difficult for me to overstate how much I value the friendship and support of people like Christopher Shaw. This is a scientist with 227 publications, most of them peer-reviewed articles. He's a tenured professor at the University of British Columbia with a distinguished career and an unmistakable moral compass. I am speaking about what I believe to be true and invite people to engage with me on these subjects. I am passionate about the problems, but am objective about both the facts and the processes used to determine what is true. I am genuinely puzzled by those who claim that to engage in a debate is equivalent to spreading misinformation.

Shaw has done extensive animal research. The importance of it from my perspective is the "cytokine storm and dysregulation of the innate immune response," or as Shaw details:

> Basically, what we see is that the animals have a number of neurobehavioral differences compared to control animals. The differences stratify by sex and age. When you take blood samples there are a whole bunch of cytokines that are massively affected. The main one is IL-5, which is interesting. [IL-5 is an inflammatory marker and was first discovered by Ruscetti and Gootenberg and called eosinophilia derived growth factor. Andy Wakefield's 1999 paper showed non-specific inflammation, eosinophilia in the guts of kids who later developed autism from an MMR shot. Yes, we know the MMR shot does not contain aluminum.]
>
> About two weeks after the animal is injected there's a 300 percent increase in IL-5. IL-5 is a microglial activator. [Microglia account for 10-15 percent of the cells in the brain and spinal cord and act as the first line of immune defense in the central nervous system. They also scavenge for "junk" in the brain, such as plaques, damaged neurons, or infectious agents.] Gradually that difference disappears by the time the animals are older. But something has changed in the brain and we don't know what it is.[19]

I find this work by Shaw to be highly alarming. If one understands the microglia act as "cleaners" of the brain, the question becomes, what are we doing to them by so heightening the immune response, as shown by the IL-5 numbers?

Are the microglia pathogenically primed by the injection of all this aluminum? They might be stuck in the activated ambeoid state, as I suggested in a slide I presented at Autism One. I most recently showed that slide in March 2021 at the Medical Academy of Special Needs (MAPS) conference as we considered the impact of the dangers of syncytin by way of COVID vaccines into these vulnerable, pathogenically primed populations.

I highly encourage everybody to get and read the books written by Stephanie Seneff and Christopher Shaw.

* * *

The great gift of Mikki's film, *Plandemic: Indoctornation* and Dr. David Martin's brilliance was it showed the world the vaccines, the mandates, and the media push behind it all was Big Pharma, Bill Gates, the World Economic Forum, and the World Health Organization. The vaccine push with the mandates is the pointy edge of the spear and there's a lot more behind it. When Shaw read our book, *Plague of Corruption,* and saw the twenty-six-minute video interview of me, "Plandemic," his suspicions were further heightened:

> You just can't understand vaccinations and autism without looking at the bigger picture . . . One part is the regulatory capture [the revolving door of scientists working for public health and Big Pharma companies] which exists in nearly every country. The role of the World Health Organization and our future overlord, Bill Gates, looms large in all of this. When you say we're not going back to normal until everybody is vaccinated, as Bill Gates has said, or that African farmers have to buy the seeds for their crops from Big Agriculture, then the bigger picture begins to peek out. The stuff you and Judy wrote about with Tony Fauci, and the way the vaccines are contaminated with animal viruses, it all started to come together.
>
> You're really looking at a multi-headed monster that has managed to seize control of health on this planet. It was never enough before, but they have to get more. When you're looking at COVID you see their end game, or at least their end game so far. It's basically to control the health of the planet and to make big bucks as well. Because the wealth of the billionaires and their friends are not going down during this pandemic. It's going up. These people are concentrating wealth and power to levels where they get to dictate what everybody does. And the politicians are either ignorant or bought off.[20]

I wish I didn't share the dark views of Chris Shaw, but I do. These people do intend harm. If there was nothing to worry about, they could easily show us the data.

We would listen and have a discussion.

However, we know we will have to force the conversation.

* * *

I think one of the most unfair portraits painted of me in the media is that I don't like to work with people. Put me with good, honest researchers and we'll get along fine. That doesn't mean we won't have lively arguments about interpretation of data, as Frank Ruscetti and I have done for thirty-eight years. No, I don't suffer fools lightly and when data doesn't support conclusions and the health of millions of innocent people are at stake, I'm not going to remain silent. I didn't serve in the military, but I did spend twenty years at the National Cancer Institute and considered that to be the most honorable of public services. When I see rampant corruption driven by a few individuals like Robert Gallo, Anthony Fauci, Harold Varmus, Ian Lipkin, and John Coffin, it hurts all the scholars who have no idea what's been done to corrupt the knowledge so carefully constructed.

All these considerations make Shaw more determined than ever to fight:

> I grew up in that fifties and sixties generation and my dad and uncles all fought in the Second World War. And you always asked yourself, if I'd been in France when the Nazis invaded, would I have joined the Maquis [rural guerilla bands of French Resistance fighters] and fought the Nazis? Everybody thinks they would have. But you know what? Most people wouldn't. Most people would just go along. It falls upon relatively few people to say this is bullshit. And I'm not going to put up with this and I'm going to resist.[21]

I guess I'm one of those people who stands up. And if you're reading this book, you're at least considering the possibility.

Shaw shares my rationale in how he will behave in this time:

> The last few years have launched me into seeing the bigger picture and to try to educate. Not to proselytize, but to educate. And you can listen to me and believe what I say or don't. But if I'm silent I haven't done my duty. I have little children and a couple of big ones and I want to be able to look them in the eye and say I did what I had to do? Did I succeed? I hope. But if I didn't, I still did what I had to do. Because how do I raise good and moral people if I don't live up to those standards?[22]

And it's good to know that, like me, Shaw is an optimist:

> Fear is very contagious. But courage is also contagious. People develop courage by seeing it in action. Which is why it's important when somebody like Bobby Kennedy stands up. With his family background, he knows it can all end very badly. But he gets up there, all the time, speaks the truth, and put himself at risk. When you watch him be brave, day in and day out, with all of his family history, you say to yourself, maybe I can show a little more courage.[23]

We need more people in science like Dr. Chris Shaw.

* * *

Dr. Frank Shallenberger first became aware of me when he watched the twenty-six-minute "Plandemic" video put together by Mikki Willis. He then bought and read our book, *Plague of Corruption*, and said to himself, "My goodness! I have such respect for Judy sticking her neck out, just so she could get her head chopped off!"[24] He was ignorant of many things we covered in our book, but felt he also had several important pieces to add to the puzzle.

I was introduced to Dr. Shallenberger by Dr. Rich Fischer during the International Academy of Oral Medicine and Toxicology conference (AMOT.org), which he had invited me to speak at in September 2020. We emailed and talked through the fall and winter of 2020. In November he invited me to attend the annual meeting of ozone therapy (AAOT.us) to be held in Dallas, Texas, May 13–15, 2021, with the single request that I commit to appear in person. I was honored and optimistic, but apprehensive as I knew nothing about ozone therapy and viruses. I did know Dr. Lori had used it on me during life-saving oral surgery.

This was the fifth medical conference I had been invited to attend live since *Plandemic: Indoctornation* had aired on August 18, 2020. I was encouraged and enthusiastic. I was being asked by experienced physicians and scientists to talk about mechanisms of pathogenesis of gamma retroviruses and vaccine contaminants. The most important thing about medical conferences is interactions with health-care practitioners and scientists in attendance. So much happens in the discussions, at dinner, coffee breaks, and at the exhibits.

Those lessons Frank Ruscetti instilled were now instinct, or as he might call it, intuition. There are no accidents. Constructing knowledge is hard work. The clear message was to study Shallenberger's work. I added Dr. Frank Shallenberger to Kent's short list to interview for this book.

As Shallenberger later recalled in a long interview with my coauthor, Kent Heckenlively:

> In the early part of my career I worked in emergency medicine and that's all I did. That and surgery also share some similarities. They're kind of unique in medicine. There's an analogy I often tell people. If somebody walks into an emergency room with a knife sticking out of their back, we actually take the knife out and sew up the wound. I joke that if somebody walked into an internal medicine office with a knife sticking out of their back, they'd give

him some Prozac to deal with the repercussions of having a knife stuck in their back. That's my complaint with internal medicine. They just don't look for causes. I did emergency medicine for five years and that's just where my head is. Getting to the cause of what's wrong.[25]

After doing emergency medicine for several years, Shallenberger started at a more traditional medical practice in internal medicine at a hospital. But after about six months, he found himself questioning his decision.

Shallenberger told the chief of the medical staff:

"Look, I'm kind of frustrated. I've only been doing this for about six months. But most of my patients, although they usually feel better, don't actually get well, unless they accidentally get well. And most of them, they're getting side-effects from the drugs I'm giving them. I don't think I'm doing a very good job here."

To his credit, the chief of the medical staff sat and discussed the issues with Frank for several hours in his office, acknowledging his concerns and finally saying, "Frank, you're doing a great job."

Shallenberger replied, "That's not satisfying."

The chief shrugged and said, "Yeah, but with chronic problems, relieving their symptoms is usually the best we can do."[26]

Around the time the chief of the medical staff told him he was doing a "great job," Shallenberger got a call from his father, also a physician, who was on the verge of retiring. While cleaning out his office, he'd stumbled across several old medical textbooks and wondered if Frank would like them.

Shallenberger later recalled:

I thought it might be fun to see what they were doing back in the early 1900s. So, he sent me all these books and I pulled out one that had been published by a large pharmaceutical company in the 1920s. I opened the first page and it was all about vitamins. I didn't realize that back in the 1920s vitamins were the big thing. I'm looking through the book and one of the very first entries was about vitamin B-1, thiamine. And they're talking about the symptoms of vitamin B-1 deficiency, things like fatigue, headaches, rashes, and lack of energy. I'm looking at this information and realizing it sounds like most of the patients I see. I'm thinking this might have something to do with B complex vitamins. I gave B complex vitamins to my patients, and dang it, a lot of them got better. The headaches really went away. And I'm thinking to myself, if somebody has a symptom, like a headache, and you give them something

natural, like a vitamin, and the symptom goes away, you probably just treated
the cause of the symptom. At that point I thought, they didn't teach me
everything I needed to know in medical school.[27]

When I read Kent's transcript of the interview, I thought of the book I'd
been given by Upjohn in 1987 called *A Century of Caring*. It described many
natural product elixirs and how one of the Upjohn brothers invented a
pill-making machine.

With this newfound appreciation that there might be answers found in
therapies pushed out of the traditional medical system, Shallenberger started
to look for fellow renegades outside of the medical establishment. He found
a study group led by two-time Nobel Prize winner Linus Pauling. Pauling
received the Nobel Prize for Chemistry in 1954 and Nobel Peace Prize in 1962,
the only scientist to receive two unshared Nobel Prizes. The group met once a
month in San Francisco. Shallenberger remembered that Pauling had come to
his medical school and given a talk about vitamin C, so he decided to attend.

A coincidence is that the first book I ever read after my grandfather
died was about vitamin C and cancer, by Linus Pauling. I even have an
autographed copy that I found at a garage sale.

After a few meetings, Shallenberger realized this was a group of physicians who were thinking way out of the box and he was excited to be a part
of them, just as we were excited about the "Amazing Papers Journal Club,"
in the early days of the Biological Response Modifiers program. Many of
them had been using vitamin therapies for fifteen or twenty years. After
about a year of attending meetings, Shallenberger had an epiphany. "Gosh,
there's all these chronic diseases going around and for the most part all the
medical system is doing is just treating the symptoms. As soon as I got that
through my skull, I realized I was going to spend the rest of my life trying
to figure out how this stuff works."[28]

Shallenberger believes his complete break with the traditional medical
establishment occurred sometime in the mid-1980s, when he officially and
irrevocably joined this renegade band of physicians who were trying to find
a better way. As an outgrowth of these monthly meetings, Pauling and other
luminaries founded the Orthomolecular Medical Society, which met once
a year and tried to summarize the important findings of the group for the
previous twelve months. One year, Frank was to put together the speaker
list and he agreed.

At the time, AIDS was dominating the national conversation and it wasn't
well-understood how the virus was killing people. One of the topics people

were talking about was oxidative stress, especially with the recent publication of a book called *Antioxidant Adaptation: It's Role in Free Radical Pathology* by Dr. Parris Kidd and Dr. Stephen Levine.[29] Oxidative stress is caused by the presence of too many "oxidants," and they were suggesting the use of certain molecules called "anti-oxidants" to battle this oversupply of oxidants. The book laid out the theory that most diseases were caused by this oxidative stress. Now oxidants and their accompanying free radicals are natural molecules formed in the body by biological processes, but a healthy person eliminates them at a higher rate than a sick person. One of the clear pieces of evidence from the HIV/AIDS patients, just as with almost all other patients with a chronic illness, was that their bodies were under immense oxidative stress. For some reason, these HIV/AIDS patients were not controlling these oxidants and free radicals, and this oversupply was destroying their immune systems.

Shallenberger got a call from a guy named Charlie Farr, an MD/PhD from Oklahoma, and he said, "I hear you're putting together the orthomolecular program in San Francisco and I want to talk."

"Okay," Shallenberger said, "What do you want to talk about?

Farr replied, "For the last couple years I've been taking hydrogen peroxide and I've been infusing it into people's veins. And they're getting better."

Shallenberger was skeptical. "Dr. Farr, that's hard to believe, knowing that hydrogen peroxide is an oxidant and people who are sick are almost always in oxidant stress. It doesn't make sense that they would get better. You would think they would only get worse."

But Farr said he had science, data, and case studies to back him up. Shallenberger was so intrigued by Farr's confidence that he immediately booked him to speak. Shallenberger recalled, "He came to the program and blew everybody's mind. He talked about a concept developed by a Swiss physician named Paracelsus, dating back to the 1400s. If you take a molecule and apply it to the body in tiny doses, it has an opposite effect of the same molecule at higher doses. It turns out, an oxidant, like hydrogen peroxide, given in tiny doses, actually decreases oxidant stress."[30] It was a very strange concept and was new to Shallenberger and virtually every other member of the Orthomolecular Medical Society. According to Shallenberger, "Charlie showed case after case of his patients who were chronically sick and in oxidative stress, as he was measuring it, then giving them intravenous hydrogen peroxide in the doses he'd developed. And dang it, their oxidant stress went away, and they got better."[31]

Shallenberger was intrigued by hydrogen peroxide and immediately started using it with success on many of his patients. But two years later he ran across a substance which seemed to have even greater applications. It was called ozone, which is actually three atoms of oxygen, sharing unstable electron bonds. As Shallenberger explained in his recent book, *The Ozone Miracle*:

> Thus, ozone, referred to as O3, is a molecule which consists of three oxygen atoms sharing the same amount of electrons that make two oxygen atoms stable. This means because of its extra oxygen atom, ozone is a relatively unstable molecule. This instability is why it is so powerful in the human body. Much more powerful that O2. And that's why when ozone is introduced into the human body, a remarkable thing happens. It instantly starts moving electrons around and ultimately stimulates the mitochondria in your cells to generate energy from oxygen more efficiently.[32]

Shallenberger started investigating ozone, finding it had first been discovered in the mid-1800s and was often used in the public water systems as a disinfectant, rather than chlorine.

In the early 1990s Shallenberger read a paper by Dr. Carpendale out of the University of California, San Francisco. He took samples of blood, bubbled ozone through them, and then poured this serum on cultures of HIV. He found it killed all the HIV in the cultures. This is from the abstract of the article, published in 1991 in the *Journal of Antiviral Resistance*:

> The inactivation of human immunodeficiency virus (HIV) and cytotoxic properties of ozone-treated serum and serum-supplemented media were examined. The titer of HIV suspensions in human serum was reduced in a dose-dependent manner when treated with total reacted ozone concentrations at a range of 0.5 to 3.5 micrograms/ml-1. Complete inactivation of HIV suspensions was achieved by 4.0 micrograms/ml-1 of ozone in the presence or absence of H-9 cells.[33]

"Complete inactivation of HIV suspension cultures in human cell lines?" This was remarkable information. Shallenberger wondered why he never knew of this work. As Shallenberger recounted to Kent:

> I'm looking at this data and I call him up. I end up hanging out with him and he tells me everything he's done. At this point, like most people, I'm convinced that AIDS was caused by a virus. I was thinking, 'Wow! AIDS is caused by a virus. You bubble ozone through somebody's blood, and it kills the virus. This is no-brainer. I'm going to be famous.' I developed this system

where I put a needle in one arm of a patient, and I had a pump attached to it. I developed an apparatus where the blood circulated through a container with a bunch of ozone in it. So, the blood got ozonated outside the body and ozonated blood was then pumped back into the other arm.[34]

Now you can understand why I call Shallenberger a problem-solver. I think it was that emergency room training which made him think, *if I understand a problem, I can fix it.*

But it wouldn't turn out to be that easy.

Shallenberger continued his recollections.

> I'm keeping these people on the system for a good three or four hours and I know damn well I'm oxidizing the hell out of their body. I published a study on four of these patients in 1998. These were patients with advanced AIDS. They didn't get cured. The virus didn't go away. The antibodies didn't go away. But the patients got better.
>
> And what I did discover was that this massive, huge dose of ozone I'd been giving them didn't work any better than a tiny dose of ozone. I might have been giving them say 400 milligrams of ozone in each session. It didn't work any better clinically than what I could get with 10 milligrams of ozone. I'm scratching my head and thinking, 'What's up with that?'
>
> Then I thought, 'Well, maybe it's the Paraclesus principle.' Different things happen with different doses. Then it occurred to me, maybe the disease isn't actually caused by the virus. If I'm killing the hell out of a virus, and the disease doesn't go away, maybe there's more to it than that."[35]

Shallenberger eventually published the results of his findings and his speculations in 1998[36], and when he read what I'd written, thought our views were highly compatible.

One of the things I most appreciate about Shallenberger is his willingness to go on the intellectual journey necessary for new knowledge to be constructed. This is true in all of science but becomes even more vital when we talk about the public's health.

Some of you might be saying, "Judy, this guy is saying AIDS isn't caused by the HIV virus." My answer is that is exactly what my PhD thesis said, but we did not appreciate all of the data at a molecular level and Shallenberger's work went against dogma, as did mine.

That's the great thing about science. It's not about the superstition of preconceived ideas. It's about following the data. None of us knows the ultimate truth, which is why we need to keep all the voices in the conversation, even the dissenting ones.

I now appreciate and honor the sacrifice of Peter Duesberg, the University of California, Berkeley researcher who clashed so publicly with Robert Gallo and Anthony Fauci about how it had not been shown that HIV caused AIDS. This was a decade before my PhD thesis and the CCR5 delta 32 deletion made it clear. Duesberg simply had the audacity to publicly say that HIV did not meet the classic definition of Koch's Postulates, in which an infectious agent must have the ability to cause the disease in otherwise healthy individuals, every single time.

But Gallo and Fauci weren't interested in having any such nuanced conversations.

Instead, Gallo and Fauci launched a vicious attack on Duesberg. They couldn't get him fired from UC Berkeley, but they could so completely choke his research funding that for years he didn't have the money to hire a single research assistant, run his lab, and could only go to his office, read the latest scientific findings, and maybe teach a few classes.

As with Dr. Andrew Wakefield and me, Duesberg was subjected to a campaign of vilification which would have given the agents of the Roman Inquisition new ideas in their persecution of Galileo.

Let me tell you what I believe.

I think HIV is associated with AIDS, just as XMRV is associated with chronic fatigue syndrome (ME/CFS), autism, cancer, and a host of other diseases.

But I do not believe the mechanisms of disease causation are well-understood.

And if we do not understand the mechanisms, we need to continue our efforts so that we may effectively intervene. That should be simple logic.

What do I think of the attacks launched by Gallo, Fauci, and others on Peter Duesberg?

Utterly despicable.

And I think the same of Gallo, Fauci, and others trying to develop AIDS drugs without understanding the mechanisms of disease causation and deliberately ignoring data as the CDC and FDA are doing now with COVID-19. They insured there would be millions of unnecessary deaths, and I consider those to be crimes against humanity.

For the benefit of those who closely follow the science, and wonder where I might agree and disagree with Peter Duesberg, this is what I believe at this moment in time.

First, HIV and XMRV do not meet the classic Koch's Postulates definition of a disease causing infectious agent, which when introduced into a healthy individual will always cause the disease. (Although HIV was considered causative because, in the 1980s, all those with AIDS had evidence of HIV infection. Except the women, of course.)

Second, Koch's Postulates need an update to take advantage of more than a century of scientific research, particularly in the areas of genetic and epigenetic vulnerability. Our 2009 *Science* paper satisfied the best available update, called Hill's Criteria, but that doesn't take into consideration the role of defective (incomplete) viral sequences or expression of HERV viruses and recombinants created through vaccine production or other processes.

Third, those villains who seek to destroy the work and reputations of good scientists asking reasonable questions, as Duesberg did, should be driven out of science and prosecuted for crimes against humanity.

Fourth, while there is strong evidence that these retroviruses and other viruses are associated with disease, we do not fully understand the mechanisms of disease causation.

Fifth, we need to understand these mechanisms to guide our treatment protocols, and effectively intervene to improve the lives of people.

* * *

Frank Shallenberger isn't a research scientist like me. He's a clinician on the front lines trying to make people's lives better. In that, we are united. He was considering the question Duesberg and many others were asking, namely, why were some people catching the virus and coming down with disease, and others weren't?

Shallenberger recalled a paper describing how the TH-1 and TH-2 immune systems were functioning in those at risk for contracting HIV and developing AIDS. The TH-1 and TH-2 systems do different things in our body, and Frank believes understanding this difference might be the key to solving many chronic diseases.

The suggestion of this TH1 to TH2 shift as being responsible for the devastating consequences of AIDS was first proposed by two researchers in 1993 in the journal, *Immunology Today*:

> This viewpoint proposes that an imbalance in the TH1-type and TH2-type
> responses contributes to the immune dysregulation associated with HIV

infection and/or progression to AIDS is dependent on a THI to TH2 dominance. This hypothesis is based on the authors' findings that: (1) progression to AIDS is characterized by loss of IL-2 [interleukin-2] and IFN gamma
[interferon gamma] production concomitant with increases in IL-4 and IL-10;
and (2) many seronegative, HIV-exposed individuals generate strong THI-
type responses to HIV antigens.[37]

What had been recognized was that there were subsets of functional T-cells,
defined by the cytokine to which they responded and produced. The shift is
a functional shift. If the cytokine expression of the T-cells changes, then the
downstream immune response changes. One of my post-doctoral projects
was defining, at the molecular level, the change in the response.

What we knew was that the promoter, the on/off switch of IL-2 and interferon gamma, was different by a single CpG nucleotide. We hypothesized the
difference between a cell that expressed IL-2 versus one expressing interferon
gamma was the methylation and silencing of that single CpG nucleotide.

Dr. Howard Young, an interferon expert and longtime colleague,
suggested a clever way to test the hypothesis. There were enzymes that
recognized and differentially cut methylated and non-methylated sites. One
enzyme was called SnaB1. With Dr. Young's guidance, I designed a PCR
test to determine if methylation of that single CpG nucleotide could distinguish at the molecular level a TH-1 versus TH-2 response.

If the CpG site was methylated the PCR product was present and it
would produce interferon gamma.

If it was unmethylated, we'd get IL-2, showing there was a functional
shift of T-cells from TH-1 to TH-2 status.

It worked.

We understood that even though only one in ten thousand T-cells
might be infected, there was a functional shift of the T-cells toward a TH-2
response, generating IL-4, IL-5, and IL-10, shifting the immune system
toward antibody production and away from the antiviral response. This was
a simple test to determine whether an HIV-infected individual was likely to
develop AIDS.

Frank Shallenberger recalled a paper he'd read looking at gay men who
had AIDS and those who did not: (I wondered if he saw the paper from my
PhD thesis, as I did that work in healthy, HIV-infected men.)

In this paper, they took a bunch of homosexual men and divided them in
half. Half of them had AIDS, and the other half was not diagnosed with

AIDS. Then they looked to see how their TH1 and TH2 systems were working. And the gist of the paper was that every single one of the men with AIDS had a TH1 to TH2 shift. Every single one. Then they looked at the control group and only 35 percent of them had a TH1 to TH2 shift. So, I'm looking at that and thinking, "Wow! Are you telling me that of these young and healthy males, 35 percent of them have already had a TH1 to TH2 shift? That can't be good. Are those 35 percent the next group about to develop AIDS?[38]

The AIDS heretic, Peter Duesberg, had suggested that the heavy use of recreational drugs and promiscuous gay sex, particularly anal sex, resulted in multiple immune system traumas to the body leading to the immune system collapse of the patients.

Duesberg didn't seem to be far off the mark.

As Frank Shallenberger recalled the change in his thinking:

> I started to realize, that what's going on with these diseases, whether it's AIDS or chronic hepatitis, or chronic Epstein-Barr. It's not the infectious agent itself that's the problem. The problem is that the particular person is not responding to the agent with an immune system in a proper manner. And ozone therapy is one way we can make you respond in a proper manner. When you've got all these people walking around, many with these TH-1 to TH-2 shifts, they're vulnerable. Along comes a virus to which these people, and they may never recover.
>
> They don't have the ability to control the virus. That's my theory. So, I'm going along and I'm treating people for viruses, and every single one of them gets better. I don't care what the virus is. Sometimes it's Hanta virus, or West Nile virus, or influenza. Most of the time I don't know what it is. I just know it's a clinical viral infection. Every single one of them gets better almost instantaneously when I apply ozone.[39]

This gets into the classic debate over whether it's the pathogen or the health of the organism which determines whether one develops a disease.

I think it's both.

Try to limit exposures to pathogens, but also try to make sure you have a healthy immune system to battle any microbes to which you might be exposed.

As Shallenberger further explains, "You do not want a TH-1 to TH-2 shift ever."[40] You want to maintain balance. Homeostasis. That may be one reason why antiparasitic drugs are beneficial in COVID-19, Ebola, and

HIV/AIDS. I always remember Dr. Dietrich Klinghardt telling me at a conference in Fuld, Germany on April 1, 2017, that he'd never seen a retrovirus associated disease where there was not a co-morbidity of a parasitic infection. Frank Ruscetti and I always say "pathogens rarely travel alone," citing the 2009 paper, "War and Peace Among the Microbes" by the Margolis lab.

Shallenberger is not a molecular biologist, but as a clinician he's trained to see patterns in his patients with chronic illness. Shallenberger states:

> There is always a trigger. I'm just a clinician. I'm a doctor and I don't know much about viruses. But I do know what makes people sick and well. And I can tell you it's darned unusual to see a patient come in with a chronic illness that can't tell me, you know, this all started with a particular series of events twenty years ago.
>
> The event was some kind of trauma to the body. It might have been a blood transfusion. It might have been a bad viral infection. They might have hit their head and had some concussion. Or they could have been big time into stress. All of these things can cause a TH-1 to TH-2 shift.[41]

Shallenberger also noted that some of his patients claimed a vaccine was a trigger for their chronic condition and he began to think more deeply about the question.

> I've never really liked vaccines. Something about them I just didn't like. At that point I asked myself, 'What's the definition of a successful vaccine?' It's one that induces the formation of antibodies. [That's not the same as life-long immunity.] Where do these antibodies come from? They come from the TH-2 system. And while the TH-2 system is busy making all these antibodies, it is also secreting cytokines that shift you from a TH-1 state to a TH-2 state. That is exactly what happens every time you get a vaccine. It shifts you from TH-1 to TH-2.
>
> And I thought, 'We're taking kids and adults and we're shooting them up with more and more vaccines.' And you have to wonder, 'How many of these kids are growing up with a normally functioning TH-1/TH-2 balance. How many are going to develop all manner of diseases and chronic conditions as a result of the shift?' There are going to be some pretty strong kids for whom it won't matter. But there's got to be a subset of the population for which it's going to be disastrous.[42]

Before reading *Plague of Corruption*, Dr. Shallenberger had a negative view of vaccines because of the TH-1 to TH-2 shift, but after reading the book, he told my coauthor, "Now you guys tell me, oh by the way, these vaccines are loaded with all these animal viruses! I thought, 'Oh my God! What are we doing?'"[43]

Like me, Shallenberger is hopeful that with proper therapies, people can get better, and can often be cured. He claims that when people come to him, they're usually the cases nobody else has been able to help. While it may take up to eight months, he estimates 85–90 percent of his patients experience substantial improvement, if not resolution of their symptoms. When the patient really commits to the program, Shallenberger believes the number increases.

We're all on a journey. But when I think of people like Dr. Frank Shallenberger, I know it's a journey I'm making with good thinkers who want to make the world a better place.

EPILOGUE

Doing the Right Thing in a World Gone Mad

Success is a collection of problems solved.

—I. M. Pei

I guess Robert Gallo can't help being a thief.

First, he stole credit for the discovery HTLV-1 and IL-2 from Frank Ruscetti. Then he tried to steal credit for the discovery of HIV from Luc Montagnier. Finally, in 2016, he tried to alter the history of human retrovirology. Maybe it shouldn't come as a surprise. Gallo reminds me of one of those aging bank robbers who just can't pass up a chance for one more heist.

I'm sure it must have been humiliating to have that report issued from the Office of Research Integrity in 1992 which drove him from the center of the scientific universe to the relative obscurity of the University of Maryland, where he has languished in exile like Napoleon on Elba for the last quarter century. Medical students of today may well have no idea that Gallo was the most quoted scientist of the 1980s and 1990s, the Anthony Fauci of his time.

But nobody today can doubt that a few scientists have become the effective rulers of our world, rather than the servants they were supposed to be.

I believe we are in a time of unprecedented danger for humanity.

Let me explain how Gallo, in his sunset years, tried to rewrite the history of XMRV research, robbing humanity of so much more than credit.

* * *

In November 2013, approximately four years after the publication of our groundbreaking article on XMRV and chronic fatigue syndrome (ME/CFS) in the journal, *Science*, and two years after I'd been erased from the scientific literature and textbooks, I delivered a talk called, "The Exotic Biology of Xenotropic Murine Leukemia Viruses (XMRVs)."[1]

I reviewed the literature as to how the retrovirus had first been identified in 2006 in approximately 10 percent of prostate cancer tissue samples by a team led by Joseph DeRisi from the University of California, San Francisco, and Robert Silverman from the Cleveland Clinic.[2]

I presented the data from our 2009 paper, which showed our original samples were negative for the VP-62 plasmid (molecular clone) positive for natural XMRV, but when those samples came back from the Silverman lab, they contained the plasmid.

What does that mean?

The VP-62 plasmid contamination was probably from Silverman's lab, which makes the most sense because Silverman and his research partner, J. Das Gupta, created VP-62 plasmid in their own lab, before I was ever in Nevada. Or did the retrovirus escape on its little viral legs from the Cleveland Clinic to my lab in Nevada?

I displayed the phylogenetic tree of XMRV, showing its close association to other mouse retroviruses, as well as how Adi Gazdar demonstrated in 2011 that the virus could spread through the air, a genuine nightmare scenario.[3]

I showed the work of the Gary Owens lab at the University of Virginia, identifying what I believe should be called XMRV-2, but has since been named B4RV. Additionally, it was found that the envelope proteins alone could cause the microvasculature changes and pathologies, not only in prostate cancer, but also in ME/CFS.[4]

In one of my slides Frank and I summarize these three points from the Owens and Coffin labs' paper:

1. B4 tumor cells harbor a retrovirus sharing 93% homology to XMRV-1 – we have designated it XMRV-2.
2. XMRV-2 is capable of infecting and stably integrating into the genome of a human prostate tumor cell line LNCaP which contains a loss of function deletion mutation within the RNAsel familial prostate cancer susceptibility gene-1.
3. XMRV-2 infected LNCaP cells show multiple functional changes associated with increased tumorgenicity and/or metastasis including

increased growth, altered migration and adhesion, and secretion of factors that decrease vascular SMC differentiation.[5]

This is exactly what one would expect with a newly discovered virus. There would be variants. Gary Owens had found a variant. Why did it get named B4RV? Because when he called it XMRV-2 at an XMRV meeting at the Cleveland Clinic presided over by John Coffin on November 10, 2009, it would have supported our *Science* paper one month after it was published.

XMRV would no longer be an orphan retrovirus, but would have a family attached to it.

And what of our claim that I believe is the reason we were attacked so viciously? That common lab experiments, mixing tissue from different animals could quickly produce variant replication competent retroviruses? That work was done by none other than the John Coffin lab. The article was entitled, "Generation of Multiple Replication-Competent Retroviruses through Recombination between Pre-XMRV-1 and PreXMRV-2."[6]

Coffin and his team found that you could generate new replication-competent retroviruses in ten days by the mixing of tissue from different types of animals and growing them in immortalized cell lines in fifty gallon fermenters!

XMRV was likely to be the tip of the iceberg of an entire asylum of manmade viruses.

If honest, ethical public health officials read and understood that paper, their first action afterward should have been shutting down all labs in which animal and human tissues are combined.

Scientific laboratories were likely becoming the greatest breeding ground in Earth's history for the creation of new and potentially dangerous pathogens.

Does the attack on our work make a little more sense now? We were telling the big boys of science that they were playing with very dangerous toys and needed to shut down their labs immediately. Recall in response to data showing seroconversion of lab workers, John Coffin stated on July 22, 2009, "We cannot afford to retrofit our labs to biosafety level three (BSL3) facilities," to prevent escape of recombinant mouse viruses.

You don't need the WHOLE infectious virus to cause harm. It's enough to simply have a partial genetic sequence from a virus to cause harm. In scientific terms, these are often referred to as "defective viruses," meaning they do not cause harm. If a researcher finds a partial genetic sequence from a virus, he will simply ignore it.

Yet, claiming that a partial genetic sequence from a virus cannot cause harm may be the single biggest body of data overlooked in the history of virology, especially retrovirology.

The other main part of my talk was that when the immune system falters, it also allows for infection by other viruses, as well as the expression of endogenous viruses like HERV-W, a family of viruses integrated into our DNA since the dawn of man. It has only recently been appreciated that endogenous viruses make up 8 percent of our genome, have a low level of expression, and are critical in the regulation of our innate immune responses, primarily the interferons.

We have known since HIV/AIDS that these patients have high levels of expression of endogenous viruses and other dormant viruses, like herpes viruses.

On the next slide, I listed the dangers which might be unleashed by these unsafe practices of scientific laboratories.

First, you might generate a new, replication-competent retrovirus never encountered by humans (which happens in every two-week production of attenuated viral vaccines).

Second, in an immune-compromised individual, it might allow for the expression of a previously silenced virus (an endogenous virus), which will cause a similar cytokine storm. We appreciated this back in 2004 when the expression of the HERV-W envelope in the brain and spinal cord was shown to cause a cytokine signature like multiple sclerosis (MS). This was like the signature of XMRV infection that we published in 2011 for ME/CFS, another disease characterized by inflammation of the brain and spinal cord.

Third, you might have a "defective" XMRV, which expressed only viral proteins, but those proteins, if expressed, could still cause disease.

And last, these defective viruses might affect regulatory expression of small inhibitory or siRNA, micro or miRNA, or long chain noncoding lcn-cRNA, a hypothesis Richie Shoemaker had discussed with us that he saw in ME/CFS and Lyme disease patients with extremely high levels of TGF-beta. Aberrant expression of all of these can cause imbalances in the functioning of the immune system.

The challenge then becomes how do we get all the prisoners back in their cells and fix the immune system damage unleashed by the retrovirus? In one of the slides, I listed my hypothesis as:

A family of human gamma retroviruses that selectively infects individuals with loss of function polymorphisms/mutations of the cancer (disease)

susceptibility gene RNASEL, and that infection increases abnormal properties such as growth and/or the metastatic properties of the tumor, stomal
cells, and immune cells.[7]

What I believe is that these pathogens are essentially causing "damage at
a distance," meaning they trigger an inflammatory cytokine storm which
leads to oxidative stress and disease. This inflammatory storm also impacts
mitochondrial function. I listed four important concepts in the wrap-up of
my presentation:

1. Recombination events in animal and human cells can generate
   families of infectious related gamma retroviruses.
2. Greatest concern is that they may acquire the ability to infect
   humans. (Though I knew they had. John Coffin and the misogynist
   gatekeepers of science had silenced my voice.)
3. Are XMRV sequences and proteins important in human disease
   pathogenesis?
4. Therapies to counteract environmentally induced aberrant gene
   RCR expression, inflammation, and immune dysregulation urgently
   need to be addressed.[8]

I felt in 2013 I'd done an excellent job of laying out the previous research
on XMRV, what Frank and I had contributed by associating it with chronic
fatigue syndrome (ME/CFS), our work isolating the virus and showing it
was infectious, and what other researchers had discovered, such as Gary
Owens with XMRV-2 (B4RV), and Adi Gazdar, showing how this virus
could float through the air.

That's the way science is supposed to work, right? We acknowledge
the work of those who came before us, tell the truth about our own contributions, and then praise those researchers who went further with our
findings.

None of that remotely describes what Robert Gallo tried to do to this
body of work in 2016.

* * *

In June 2016, Gallo and others published an opinion paper in the
Proceedings of the National Academy of Sciences (*PNAS*) with the curious
title, "Extracellular Vesicles and Viruses: Are They Close Relatives?"[9]

Are you thinking to yourself, oh great, "extracellular vesicles," another scientific term I don't understand? All retroviruses bud off the host cell membranes. The membranes package the RNA protecting it from degradation by a healthy immune system. Back in the 1980s, Bruce Lipton was the first to describe exosomes, shuttling regulatory RNA from cell to cell in a normal immune response.

Here's the opening paragraph of Gallo's article:

> Cell in vivo and ex vivo release membrane vesicles. These extracellular vesicles (EVs) are 50-100-nm-sized lipid bilayer-enclosed entities containing proteins and RNA. Not long ago, EVs were considered to be "cellular dust" or garbage and did not attract much attention. However, it has recently been found that EVs can have important biological functions and that in both structural and functional aspects they resemble viruses. This resemblance becomes even more evident with EVs produced by cells productively infected with viruses. Such EVs contain viral proteins and parts of viral genetic material. In this article, we emphasize the similarity between EVs and viruses, in particular retroviruses. Moreover, we emphasize that in the specific case of virus-infected cells, it's almost impossible to distinguish EVs from (noninfectious) viruses and to separate them.[10]

The final sentence of the paragraph is the real kicker, the place where the dishonesty of the man should be clear to any person who pays attention: "Moreover, we emphasize that in the specific case of virus-infected cells, it is almost impossible to distinguish EVs from (noninfectious) viruses and separate them."

That statement is simply not true.

Is Gallo going to rewrite the history of retrovirology and explain away the crimes of the past four decades? It is NOT impossible to distinguish extracellular vesicles and retroviruses.

That is why one would use an electron micrograph.

It is the gold standard to distinguish between families of retroviruses.

Electron micrographs absolutely and unequivocally distinguish viruses from exosomes and extracellular vesicles. Retroviruses and coronaviruses have envelope or spike proteins poking through the host lipid bilayer. And the core of the virus has a characteristic pattern not seen in exosomes or extracellular vesicles packaging mRNA or defective viral genomes.

Please don't fall for this latest Gallo lie.

* * *

As I mentioned before, the result of our *New York Times* bestselling book *Plague of Corruption* and the video "Plandemic" is that scholars were waking up, seeing the connections, and the possible existing solutions to the problems. After nearly a decade in exile, I was being invited again to give scientific talks at medical conferences.

And in March 2021, I was invited to give a talk to the Medical Academy of Pediatric Special Needs (MAPS) and the International Academy of Oral Medicine and Toxicology (AOMT). These were back-to-back conferences in Florida and they'd even booked me a first-class ticket. Not only would I be educating doctors, but I would be receiving an award for courage in medicine. On March 10, 2021, I had a flight booked from the Santa Barbara, California airport to Florida on Skywest Airlines (American Eagle), part of American Airlines.

I arrived at the airport an hour before my flight and wore the triple layer heavyweight silver mesh pleated mask prescribed by my doctor because I have a precancerous esophageal condition known as Barrett's esophagus. (This was probably exacerbated by XMRV infection, because like many of the lab workers, I was a control who acquired XMRV via a lab infection expressing not only antibodies but high levels of envelope protein.)

I went through security with my silver mask and boarded the plane second in line, past two gate agents.

It was odd when the flight attendant said to me, "Welcome aboard, Ms. Mikovits."

It had been a long time since I'd flown first-class, and I'd forgotten that perk. They welcome you by name.

"Thank you," I replied.

"Would you like a hand sanitizer?" the flight attendant offered.

"No, thank you," I replied. Have you ever read the list of ingredients in the typical hand sanitizer? Give me the good old-fashioned soap and water every time. Someday soon, a good products liability lawyer is going to get those products banned.

I proceeded to my seat, 2D, and began loading my luggage in the overhead bin.

The flight attendant followed behind me and said, "You cannot wear that mask."

I turned to her and replied, "Pardon me? It is an appropriate mask. It's a triple layered mesh and silver mask, prescribed to me by my doctor because I have a lung disease."

The flight attendant thrust one of those blue, chemically treated, paper masks that violate California's Prop 65 laws at me and threatened, "You must agree to wear this mask for the entire flight, or you will be removed."

"I cannot say or do that," I replied, knowing this was unlikely to go well for me. "You're not a medical professional, and there are no masks 'approved' by the FDA, CDC, or any other authority. Additionally, airlines and airports are places of public accommodation and people are guaranteed equal access, regardless of medical condition or religious belief, as well as other rights guaranteed under state and federal laws." I had a copy of our book, *The Case Against Masks: Ten Reasons Why Mask Use Should be Limited*, published in the summer of 2020, and electron micrographs of the nanofiber and microbial particulates on the slides I was to present at the conference.

Amazon had carried the book for a few weeks and then banned it, so I brought the only two copies I had with me, which I had autographed for the organizers of the meeting as a thank you gift.

By coincidence, our publisher, Skyhorse Publishing, has another book, *The Case for Masks*, which is sold on Amazon. Got it? Amazon allows the sale of a book in favor of masks but doesn't allow you to read a book suggesting there should be limits on mask use.

That's why, when that flight attendant confronted me about my mask, I knew my rights. I'd done my homework, like I always do.

I sat down in seat 2D as I continued to talk to the flight attendant, challenging her to show me any evidence that my mask wasn't acceptable. She did not produce any evidence.

The police were called by the flight attendant and the pilot.

Now, think about this situation. I am a sixty-two-year-old woman, and I was complying with what I believed to be illegal mandates by wearing a safe mask for my medical condition, which was exacerbated when I acquired a contagious cancer and ME/CFS-associated retrovirus infection doing critical research for humanity. I later came to believe this flight attendant must have known who I was, no doubt believing the lies spread by the mainstream media about me, and decided it was her day to knock down an enemy of the people.

I refused to leave, saying I had violated no laws or rules and demanded they call the sheriff, who had vowed to protect the constitutional rights of his constituents.

The police officer became excessively violent with me (I was also trying to film with my cellphone at the time) as I finally agreed to stand up and unbuckled my seat belt.

The officer them grabbed my left arm, violently twisted it behind my back, and so brutally pushed me up against the window that I screamed and saw stars. I was handcuffed in that position, and they took my cellphone. They left my mask in my handcuffed hands and marched me down the jetway, thereby implying I refused to wear a mask. I heard other passengers yelling obscenities and statements like "Throw away the key!"

I was then walked to the street where there were two waiting SUVs. They demanded I sign the citation of criminal trespass, or I'd be taken away to jail. I repeatedly refused to sign and asked them to call the sheriff so I could plead my case directly to him.

My left arm was in great pain from being brutally thrown against the window, and not wanting to spend any more time in jail than I already had in my life, I signed the citation and was released.

I gave my presentation the next day over Zoom, from my five-hundred-square-foot apartment in Ventura, California, and got generous applause when I finished.

I soon got legal representation and will take legal action against all involved in this unnecessary and violent assault of me.

This madness has got to stop.

* * *

That's my life.

I get physically assaulted in airports even while wearing a mask, denounced in the mainstream media, and attacked in print in scientific journals. And, of course, they won't allow me to publish a response. In July 2020, the journal, *AIDS Research and Human Retroviruses*, published an article with the title, "Fake Science: XMRV, COVID-19, and the Toxic Legacy of Dr. Judy Mikovits." Here's the abstract:

One cannot spend more than five minutes on social media at the moment without finding a link to some conspiracy theory or other regarding the origin of SARS-CoV2, the coronavirus responsible for the COVID-19 pandemic. From the virus being deliberately released as a bioweapon to pharmaceutical companies blocking the trials of natural remedies to boost their dangerous drugs and vaccines, the Internet is rife with far-fetched rumors. And predictably, now that the first immunization trials have started, the antivaccine lobby has latched onto most of them. In the last week, the trailer for a new "bombshell documentary" *Plandemic* has been doing the rounds, gaining notoriety for being repeatedly removed from YouTube and Facebook. We usually would not pay much heed to such things, but for retrovirologists like us, the name associated with these claims is unfortunately too familiar: Dr. Judy Mikovits.[11]

How would you feel if you woke up one day and read that about yourself as you were having your morning coffee?

Let's talk for a moment about the claims I started making in 2020 as the insanity of the COVID-19 crisis was kicking into high gear. Probably the thing that made my colleagues in public health the angriest was that I was claiming SARS-CoV-2 was most likely a release from the Wuhan Institute of Virology. And why did I say that? Because others in the scientific community had been talking about the risk of "gain of function" research for several years.

This is from an article published in the journal, *Nature*, on November 12, 2015, by Declan Butler with the title, "Engineered Bat Virus Stirs Debate Over Risky Research."

> An experiment that created a hybrid version of a bat coronavirus – one related to the virus that created SARS (severe acute respiratory syndrome) – has triggered renewed debate over whether engineering lab variants of viruses with possible pandemic potential is worth the risks.
>
> In an article published in Nature Medicine on November 9, scientists investigated a virus called SHC014, which is found in horseshoe bats in China. The researchers created a chimaeric virus, made up of a surface protein of SHC014 and the backbone of a SARS virus that had been adapted to grow in mice and to mimic human disease. The chimaera infected human airway cells – proving that the surface protein of SHC014 has the necessary structure to bind to a key receptor on the cells and to infect them.[12]

I genuinely don't understand the criticism of the scientific community when I'm simply pointing out what they've said in their own journals. I had nothing to do with this publication. I'm simply drawing people's attention to it. And how is it that something that scientists were actually debating in 2015 suddenly became the ultimate act of scientific heresy in 2020, when I, along with others, commented on it? The article continued:

> But other virologists question whether information gleaned from the experiment justifies the potential risk. Although the extent of any risk is difficult

to assess, Simon Wain-Hobson, a virologist at the Pasteur Institute in Paris, points out that the researchers have created a novel virus that "grows remarkably well" in human cells. "If this virus escaped, nobody could predict the trajectory," he says.

The argument is essentially a rerun of the debate over whether to allow lab research that increases the virulence, ease of spread or host range of dangerous pathogens – what is known as 'gain of function' research. In October 2014, the US government imposed a moratorium on federal funding of such research on the viruses that cause SARS, influenza, and MERS (Middle East respiratory syndrome, a deadly disease caused by a virus that sporadically jumps from camels to people.)

The latest study was already under way before the US moratorium began, and the US National Institutes of Health (NIH) [actually the National Institute of Allergy and Infectious Diseases, headed by Dr. Anthony Fauci] allowed it to proceed while it was under review by the agency, says Ralph Baric, an infectious-disease researcher at the University of North Carolina at Chapel Hill, a coauthor of the study. The NIH eventually concluded that the work was not so risky as to fall under the moratorium, he says.[13]

I don't know about you, but the conclusion that such work was "not so risky" might be the biggest miscalculation in history since the US Navy decided their fleet at Pearl Harbor wasn't at any risk from a Japanese attack.

Remember that all through 2020 I was attacked for saying that the virus likely escaped from the Wuhan Institute of Virology.

On March 26, 2021, the former director of the CDC, Dr. Robert Redfield, finally caught up to my opinion in an interview with CNN's Dr. Sanjay Gupta, and covered in an article titled, "Former CDC Director Surprises CNN's Sanjay Gupta by Revealing He Believes COVID-19 Originated in a Wuhan Lab." From the opening of the article:

> CNN's Sanjay Gupta appeared taken aback in a new interview as former CDC Director Robert Redfield shared his "opinion" on the origins of COVID-19.
>
> Gupta spoke with Redfield as part of a new CNN documentary about the COVID-19 pandemic, and the former CDC chief revealed he thinks it's likely the coronavirus originated in a Wuhan lab and was possibly being transmitted in September 2019.
>
> "If I was to guess, this virus started transmitting somewhere in September, October in Wuhan," Redfield said. "That's my own view. It's only an opinion. I'm allowed to have opinions now."

Redfield continued that he thinks the "most likely etiology of this pathogen in Wuhan was from a laboratory, you know, escaped," though he said he's not "implying any intentionality" and reiterated "it's my opinion."[14]

Is it just me, or when you read that section does it seem like the former director of the CDC is pathetically pleading not to be the next victim of cancel culture, or worse yet, vilified by the powers that be, and completely lose his status?

Redfield knew, said, and did nothing when he worked on Trump's task force, and then lied for a year while millions of lives were ruined. He also coauthored a February 1, 2020 article with Anthony Fauci saying that masks don't work.

I wish I could come up with some way to show my supreme disrespect for this kind of cowardice, but I think I'll just let you use your imagination.

* * *

The average person might reasonably ask, what is to be done?

I'm telling you what I'm doing. I'm trying to get to the bottom of this controversy. I don't think that China is solely to blame for this problem, but the United States is as well. It doesn't matter where the viruses originally came from. Immortalized cell lines containing many unsuspected viruses have been shipped around the world for more than a quarter of a century.

We need to STOP all this research and declare a moratorium on all vaccinations on the schedule until the animal and human aborted fetal tissue is removed and the double-blind placebo-controlled studies are done.

The next step is to remove the serial felons and liars from the public health organizations and restore liability and informed consent for all pharmaceuticals, including vaccines.

What I've tried to convey in this book is that there's a scientific elite that exists above presidents and prime ministers, and they genuinely refuse to reveal the truth of what they're doing in their labs, even to their own leaders. The president of China is probably in the dark just as much as our own leaders about what actually happened. I have tried to make this transparent to the general public by writing these books because I believe we, as scientists, are public servants, and you, the general public, deserve to know what's happening in these labs.

The spirit in which I write this book is in accord with the declaration by President Abraham Lincoln during the darkest days of our Civil War that: "I am a firm believer in the people. If given the truth, they can be depended upon to meet any national crisis. The great point is to bring them the real facts."

Toward that end, with the help of attorney Larry Klayman and his organization, Judicial Watch, I have joined their class action as an expert witness against the government of China, the People's Liberation Army, and the Wuhan Institute of Virology for their role in the COVID-19 crisis and to discover the truth. From the opening of the action which was filed in the United States District Court for the Northern District of Texas on March 17, 2020:

1. This is a complaint for damages and equitable relief arising out of the creation and release, accidental or otherwise, of a variation of coronavirus known as COVID-19 by the People's Republic of China and its agencies and officials as a biological weapon in violation of China's agreement under international treaties, and recklessly or otherwise allowing its release from the Wuhan Institute of Virology into the city of Wuhan, China, in Hubei Province, by among other acts failing to prevent the Institute's personnel from becoming infected with the bioweapon and carrying it into the surrounding community and proliferation into the United States.

2. Since biological weapons have been outlawed since at least 1925, including by China's membership in treaties, these illegal weapons constitute and are in effect terrorist-related weapons of mass destruction of population centers.[15]

Many have warned me against taking such bold, public action, but I am a woman of faith and fear only the Lord. We are not promised a life free of suffering. I look forward to a life in eternity, hearing the words, "Well done, good and faithful servant." We all deserve the truth from our leaders and the best health that science can give us. We have a God given immune system that can defend us from disease, no matter what is created in laboratories. The Lord has given us a great abundance on this earth to keep us healthy. He has not left us defenseless in this time of terror.

It's up to each one of us to maintain the purity of our body temples, as created by God. The mark of the beast is not the needle in the arm. It's the fear in the mind that causes us to doubt.

We have all been deceived. But each of us can repent and change our hearts and minds, never allowing ourselves to fall into the trap of fear and hatred like COVID-19 again.

Forgive yourself and forgive others.

The obligation of a scholar, and indeed every citizen, is to discover the truth, speak the truth, and live in truth. I hope that through my efforts and example I have helped shed light on some very dark places in science and encouraged a healthy resurgence of true, unbiased science.

If I have brought light to these issues, then perhaps I will have enhanced my fervent wish that the best of health finds its way to all of us.

Notes

PART ONE
Chapter Two

1 F.W. Ruscetti and P.A. Chervenick, "Release of Colony Stimulating factor from monocytes by Endotoxin and Poyino-Sinic-Polycytidic Acid," *Journal of Laboratory and Clinical Medicine* 83 (January 1974) 64-72.
2 F.W. Ruscetti, P.A. Chevernick, "The Release of Colony-Stimulating Factor from Thymus-Derived Lymphocytes," *Journal of Laboratory and Clinical investigation* 55 (March 1975) 520-527.
3 F.W. Ruscetti, R. Cypress, and P.A. Chevernick, "Specific release of Neutrophil and Eosinophil Stimulating Factors from Sensitized Lymphocytes," *Blood* 47 (May 1976) 757-765.
4 Norman Wolmark, "In Memoriam: Bernard Fisher, 1918-2019," *Journal of Clinical Oncology*, 38. No. 16 (April 14, 2020), 1751-1756, www.doi.org/.1200/JCO.19.03299.

Chapter Three

1 Nicholas Wade, "Special Virus Cancer Program: Travails of a Biological Moon Shot," *Science*, no. 4106 (December 24, 1971): 1306-1311.
2 R.E. Gallagher and R. Gallo, "Type C RNA Tumor Virus Isolated from Cultured Human Acute Myelogous leukemia Cells," *Science*, 187 (January 31, 1975): 350-353, www.doi.org/1o.1126/sceince.46123.
3 N.M. Teich, R.A. Weiss, S.Z. Salahuddin, R.E. Gallagher, and R.C. Gallo, "Infective Transmission and Characterisation of a C-Type Virus Release by Cultured Human Myeloid Leukemia Cells," *Nature* 256 (5518), (August 14, 1975): 55-1-555, www.doi.org/10.1038/256551a0.
4 D.A. Morgan, F.W. Ruscetti, and R.C. Gallo, "Growth of Thymus-Derived Lymphocytes from Normal Human Bone Marrow," *Science* 193 (1976) 1007-1008, F.W. Ruscetti, D.A. Morgan, and R.C. Gallo, "Functional and Morphological Characteristics of Human Thymus-Derived Lymphocytes Continuously Growing in Vitro," *Journal of Immunology* 119 (July 1977) 131-138,

5   R. Gallo, W. Saxinger, R. Gallagher, et al., "Some Ideas on the Origin of Human
     Leukemia in Man and Recent Evidence for the Presence of Type-C Related Viral
     Information," *Origins of Human Cancer*, Cold Spring Press (Cold Springs Harbor, New
     York - 1977), p. 1253-1265.

6   S. Gillis, P. E. Baker, F. W. Ruscetti, and K. A. Smith, "Long Term Culture of Human
     Antigen-Specific Cytotoxic T-Cell Lines," *Journal of Experimental Medicine*, Vol. 148 (4),
     pp. 1093-1098: doi:10.1084/jem.148.4.1093.

7   Ibid.

8   Ibid at 1093.

9   Steve Rosenberg, "IL-2: The First Effective Immunotherapy for Human Cancer," *Journal
     of Immunology*, Vol. 192, (12), pp. 5451-5458, 5451: doi: 10.4049/jimmunol.1490119.

10   American Association of Immunology Meeting in Honolulu, Hawaii, www.drive.google
     .com/file/d/1CS7MV5q3BoHz-4usxe4HlNsVdrK-1g1J/view

11   S. Collins, F. Ruscetti, F. Gallagher, and R. Gallo, "Terminal Differentiation of
     Human Promyelocytic Leukeima Cells Induced by Dimethyl Sulfide and Other Polar
     Compounds," *Proceedings of the National Academy of Sciences* 75 (May 1978) 2458-2462,
     www.doi.org/10.1073.pnas.75.5.2458.

12   T.R. Breitman, Stuart E. Selonick, and Steven J. Collins, "Induction of Differentiation of
     the Human Promyelocytic Leukemia Cell Line (HL-60) by Retinoic Acid," *Proceedings
     of the National Academy of Sciences* 77, no. 5 May 1980): 2936-2940.

13   "Blood Flashback – 1979: Characterization of the Continuous Differentiating Myeloid
     Cell Line (HL-60) From a Patient with Acute Promyelotic Leukemia," *American Society
     of Hematology*, (2016): www.doi.org/10.1182/blood-2016-10-748780.

14   A.F. Gazdar, D.N. Carney. P.A. Bunn, E.K. Russel, E.S. Jaffe, G.P. Schechter, and J.G.
     Guccion, "Mitogen Requirements for the In Vitro Propogation of Cutaneous T-Cell
     Lymphomas," *Blood* 55, no. 3 (March 1980): 409-417.

15   B. Poiesz, F.W. Ruscetti, A. Gazdar, et al., "Isolation of Type-C Retrovirus particles from
     Cultured and Fresh Lymphocytes from a Patient with Cutaneous T-Cell Lymphoma,"
     Proceedings of the National Academy of Sciences 77 (December 1980) 7415-7419.
     www.doi.org/10.1073/pnas.77.12.7415.

16   John M. Coffin, "The Discovery of HTLV-1, the First Pathogenic Human Retrovirus,"
     Proceedings of the National Academy of Sciences 51, 15525 – 15529, www.doi.org/10.1073
     /pnas.1521629112.

17   S.Z. Salahuddin, P.D. Markham, F.W. Ruscetti, R.C. Gallo, "Long-Term Suspension
     Cultures of Human Cord Blood Myeloid Cells," *Blood* 58 (November 1981) 931-938,
     www.doi.org/10.1182/blood.V58.5.931.931.

18   J.E. Gootenberg, F.W. Ruscetti, J.W. Mier, et al., "Human Cutaneous T-Cell Lymphoma
     Lines Produce and Respond to T-Cell Growth Factor," *Journal of Experimental Medicine*
     154 (November 1981) 1403-1418, www.doi.org/10.1084/jem.154.5.1403.

19   J.E. Gootenberg, F.W. Ruscetti, R.C. Gallo, "A Biochemical Variant of T-Cell Growth
     Factor from a Human T-Cell Lymphoma Line," *Journal of Immunology* 129 (October
     1982) 1499-1505.

Chapter Four

1   Robert K. Merton, "The Matthew Effect in Science," *Science* 159, 56-63 (January 5,
     1968): 56-63. www.doi.org/10.1126/science.159.3810.56.

2   Nicholas Wade, "Special Virus Cancer Program: Travails of a Biological Moon Shot," *Science*, no. 4106 (December 24, 1971): 1306-1311.

Chapter Five

1   Aldous Huxley, *Complete Essays, Vol. II: 1926-1929* (Chicago: Ivan Dee, 2000), (Notes on Dogma)
2   *Macbeth*, Act 3, Scene 2, Page 3.
3   L. Wolff and S. Ruscetti, "Malignant Transformation of Erythroid Cells In Vivo by Introduction of a Non-Replicating Retrovirus Vector," *Science* 228 (June 28, 1985): 1549-1552, www.doi.org/10.1126/science.2990034.
4   F. Barré-Sinoussi, J.C. Chermann, F. Rey, M.T. Nugeyre, S. Chamaret, J. Gruest, C. Daugert, C-Axler-Blin, F. Vezinet-Brun, C. Rouziox, W. Rosenbaum, and L. Montagnier, "Isolation of a T-Lymphotropic Retrovirus from a Patient at Risk for Acquired Immune Deficiency Syndrome (AIDS)," *Science* 22, (May 20, 1983): 868-871, www.doi.org/10.1126/science.6189183.
5   R.C. Gallo, P.S. Sarin, E.P. Gelmann, M. Robert-Guroff, E. Richardson, V.S. Kalyanaraman, D. Mann, G.D. Sidhu, R.E. Stahl, S. Zolla-Pazner, J. Leibowitch, and M. Popovic, "Isolation of Human T-Cell leukemia Virus in Acquired Immune Deficiency Syndrome (AIDS)," *Science* 220 (May 20, 1983): 865-867, www.doi.org/10.1126/science.6601823.
6   J.A. Levy, A.D. Hoffman, S.M. Kramer, J.A. Landis, J.M. Shimabukuro, and L.S. Oshiro, "Isolation of Lymphocytopathic Retroviruses from San Francisco Patients with AIDS," *Science* 225 (August 24, 1984): 840-842, www.doi.org/10.1126/science.6206563.
7   R.C. Gallo, P.S. Sarin, E.P. Gelmann, M. Robert-Guroff, E. Richardson, V.S. Kalyanaraman, D. Mann, G.D. Sidhu, R.E. Stahl, S. Zolla-Pazner, J. Leibowitch, and M. Popovic, "Isolation of Human T-Cell leukemia Virus in Acquired Immune Deficiency Syndrome (AIDS)," *Science* 220 (May 20, 1983): 865-867, www.doi.org/10.1126/science.6601823.
8   Serge Lang. "The Gallo Case," (1989), pp. 364 – 632, 585 www.link.springer.com/content/pdf/10.1007%2F978-1-4612-1638-4_5.pdf.
9   Ibid. at p. 490.
10   Philip J. Hilts, "Federal Inquiry Finds Misconduct by a Discoverer of the AIDS Virus," *New York Times*, December 31, 1992, www.nytimes.com/1992/12/31/us/federal-inquiry-finds-misconduct-by-a-discoverer-of-the-aids-virus.html.

Chapter Six

1   Society for Immunotherapy of Cancer, "Leading Cancer Immunotherapy Scientists and Teams," October 4, 2010, www.sitcancer.org/aboutsitc/press-releases/2010/leading-scientists.
2   Austin Frakt, "Reagan, Deregulation and America's Exceptional Rise in Health Care Costs," *New York Times*, June 4, 2018, www.nytimes.com/2018/06/04/upshot/reagan-deregulation-and-americas-exceptional-rise-in-health-care-costs.html
3   Olivia Campbell, "Here's What Happened when Reagan Went After Healthcare Programs. It's Not Good," *Timeline*, September 13, 2017, www.timeline.com/reagan-trump-healthcare-cuts-8cf64aa242eb.

⁴ Ibid.

⁵ Ibid.

⁶ Ray Sipherd, "The third leading Cause of Death in US Most Doctors Don't Want to Talk About," *CNBC*, February 22, 2018, www.cnbc.com/2018/02/22/medical-errors-third -leading-cause-of-death-in-america.html

⁷ Leah Sittile, "Living Sick and Dying Young in Rich America," *The Atlantic*, December 19, 2013, www.theatlantic.com/health/archive/2013/12/living-sick-and-dying-young -in-rich-america/282495/

⁸ Alan L. Hillman, David B. Nash, William L. Kissick, and Samuel P. martin, "Managing the Medical Industrial Complex," *New England Journal of Medicine* 315 (August 21, 1986): 511-513, www.doi.org/10.1056/NEJM198608213150810.

⁹ Alexader Zaitchik, "How Big Pharma was Captured by the One Percent," *The New Republic*, June 28, 2018, www.newrepublic.com/article/149438/big-pharma-captured -one-percent

¹⁰ Ibid.

¹¹ Ibid.

¹² Andrew Pollack and Michael. J. de la Merced, "Gilead to Buy Pharmasset for $11 Billion," *The New York Times*, November 21, 2011, www.dealbook.nytimes.com/2011/11/21 /gilead-to-buy-pharmasset-for-11-billion/

¹³ P. Gosselin, "How the $250 Billion US Diabetes Industry Operates . . . There's a Lot of money to be Made Keeping You Sick," *No Tricks Zone*, January 2, 2016, www.notrickszone.com/2016/01/02/how-the-250-billion-us-diabetes-industry -operates-theres-a-lot-of-money-to-be-made-keeping-you-sick/

¹⁴ Nora D. Volkow, "The Biology and Potential Therapeutic Effects of Cannabidol," *National Institute on Drug Abuse*, June 24, 2015, www.archives.drugabuse.gov /testimonies/2015/biology-potential-therapeutic-effects-cannabidiol

¹⁵ Arnold Relman, Marcia Angel, "America's Other Drug Problem," *New Republic*, (December 12, 2002), www.newrepublic.com/article/66623/americas-other-drug -problem

¹⁶ Marcia Angell, "Drug Companies & Doctors: A Story of Corruption," *The New York Review of Books*, January 15, 2009, www.nybooks.com/articles/2009/01/15/drug -companies-doctorsa-story-of-corruption/

¹⁷ "Opioid Overdose Tops List of Leading Causes of Death in the U.S." *American Society of Hematological Clinical News*, January 18, 2019, www.ashclinicalnews.org/online -exclusives/opioid-overdose-tops-list-leading-causes-death-u-s/

¹⁸ Robert Pear, "Reagan Signs Bill on Drug Exports and Payment for Vaccine Injuries," *New York Times*, November 15, 1986, www.nytimes.com/1986/11/15/us/reagan-signs -bill-on-drug-exports-and-payment-for-vaccine-injuries.html

¹⁹ Jacob Hornberger, "The CIA's Longstanding relationship with Crime," *Future of Freedom Foundation*, August 13, 2020, www.fff.org/2020/08/13/the-cias-longstanding -relationship-with-crime/

²⁰ Landmark Recovery Staff, "The History of the War on Drugs: Reagan Era and Beyond," *Landmark Recovery*, February 13, 2019, www.landmarkrecovery.com/history-of-the -war-on-drugs-reagan-beyond/

²¹ Harvey J. Alter and Harvey G. Klein, "The Hazards of Blood Transfusion in Historical Perspective," *Blood* 112 (October 1, 2008): 2617-2626, www.doi.org/10.1182 /blood-2008-07-077370.

22 P.M. Hoffman, B.W. Festoff, L.T. Giron, Jr., L.C. Hollenbeck, R.M. Garruto, and
    F.W. Ruscetti, "Isolation of LAV/HTLV-III from a Patient with Amyotrophic Lateral
    Sclerosis," *New England Journal of Medicine* 313 (August 1, 1985): 324-325, www.doi
    .org/10.1056/NEJM198508013130511.

23 J.L. Marx, "Lasker Award Stirs Controversy," *Science* 203, (January 26, 1979): 341,
    www.doi.org/10.1126/sicence.216074.

24 Paul Trachtman, "Review of 'Molecules of Emotion,' book by Candace Pert,"
    Smithsonian Magazine, September 1998, www.smithsonianmag.com/arts-culture
    /review-of-molecules-of-emotion-157256854/#:~:text=In%20Molecules%20of%20
    Emotion%2C%20Pert,obscures%20the%20pursuit%20of%20truth.

25 Ibid.

26 Maria T. Polianova, Francis W. Ruscetti, Candace B. Pert, and Michael Ruff, "Chemokine
    receptor-5 (CCR5) is a receptor for the HIV Entry Inhibitor Peptide T (DAPTA),
    Journal of Antiviral Resistance, 67 (August 2005): 83-92, www.doi.org/10.1016/j
    .antiviral.2005.03.007.

27 Hyeryun Choe, Michael Farzan, Ying Sun, Norma Gerard, raig Gerard, and Joseph
    Sodroski, "The B-Cemokine Receptors CCR3 and CCR5 Facilitate Infection by Primary
    HIV-1 Isolates," *Cell* 85 (June 28, 1996), 1135-1148, www.doi.org/10.1016/S0092
    -8674(00)81313-6.

28 Julie Catusse, Chris M. parry, David R. Dewin, and Ursula A. Gompels, "Inhibition
    of HIV-1 Infection by Viral Chemokine U83A via High Affinity CCR5 Interactions
    that Block Human Chemokine-Induced Leukocyte Chemotaxis and Receptor
    Internalization," *Blood* 109 (January 5, 2007): 3633-3639, www.doi.org/10.1182
    /blood 2006-00-042622.

29 Mary T. Joy, Einor Ben Assayag, Dalia Shabashov-Stone, et al., "CCR5 is a Therapeutic
    Target for Recovery after Stroke and Traumatic Brain Injury," *Cell* 176 (February 21,
    2019); 1143-1157, www.doi.org/10.1016/j.cll.2019.01.044.

30 "Side Effects of Maraviroc," *Clinical Information, HIV GOV* (Last reviewed on November
    18, 2020), www.clinicalinfo.hiv.gov/en/drugs/maraviroc/patient

Chapter Seven

1   "Frans De Wahl," "GoodReads," (Accessed July 16, 2021), www.goodreads.com
    /quotes/760359-the-enemy-of-science-is-not-religion-the-true

2   "Brainy Quotes – John Maynard Keynes" (Accessed July 16, 2021, www.brainyquote.
    com/quotes/john_maynard_keynes_385471

3   F.W. Ruscetti, L. Varesio, A. Ochoam and J. Ortaldo, "Pleiotropic Effects of Transforming
    Growth factor-B on cells of the Immune System," *New York Academy of Sciences* 685
    (1993): 488-500, www.doi.org/10.1111/j.1749-6632.1993.tb35911.x, and F. Ruscetti,
    M. Birchenall-Roberts, J. McPherson and R. Wiltrout, "Transforming Growth Factor-B.
    In. Cytokines, A. Sluis (Ed), (Academic Press, New York, New York) 1998, 415-432.

4   "HIV and the Blood Supply: An Analysis of Crisis Decision Making," *Institute of Medicine*,
    July 13, 1995, www.drive.google.com/file/d/1jkVLXlrOZTzgx37RcobYgv9GfsG-5sbB
    /view

5   Robert Pear, "AIDS Blood Test to be Available in 2 to 6 Weeks," *New York Times*,
    March 3, 1985, www.nytimes.com/1985/03/03/us/aids-blood-test-to-be-available-in-2
    -to-6-weeks.html

6   "How One Test Changed HIV," *Abbott Labs Media*, November 27, 2019, www.abbott
     .com/corpnewsroom/products-and-innovation/how-one-test-changed-HIV.html

7   "The Inadequate Response of the FDA to the Crisis of AIDS in the Blood Supply,"
     Harvard Library Office of Scholarly Communication (1995), www.dash.harvard.edu
     /handle/1/8965576

8   Walt Bogdanich and Eric Koll, "2 Paths of Bayer Drug in 80's: Riskier One Steered
     Overseas," *New York Times*, May 2, 2003, www.nytimes.com/2003/05/22/business/2
     -paths-of-bayer-drug-in-80-s-riskier-one-steered-overseas.html

9   Ibid.

10   "HIV and the Blood Supply: An Analysis of Crisis Decision Making," *Institute of Medicine*,
     July 13, 1995, www.drive.google.com/file/d/1jkVLXlrOZTzgx37RcobYgv9GfsG-5sbB
     /view

11   Ibid.

12   Celia Farber, "AIDS and the AZT Scandal," *Spin Magazine* (November 1989), www
     .spin.com/featured/aids-and-the-azt-scandal-spin-1989-feature-sins-of-omission/

13   Alice Park, "The Story Behind the First AIDS Drug," *TIME*, March 19, 2017, www
     .time.com/4705809/first-aids-drug-azt/

14   "Exhibition – Fight Back, Fight AIDS," *U.S. National Library of Medicine*, (Accessed
     July 16, 2021). www.nlm.nih.gov/exhibition/survivingandthriving/exhibition-fight
     -back-fight-aids.html

15   "Women and HIV: A Spotlight on Adolescent Girls and Young Women," *UN AIDS*,
     (2019), www.unaids.org/sites/default/files/media_asset/2019_women-and-hiv_en.pdf

16   J.A. Mikovits, Raziuddin, M. Ruta, et al., "Negative Regulation of HIV Replication
     in Monocytes: Distinctions Between Restricted and Latent Expression in THP-1
     Cells," *Journal of Experimental Medicine* 171 (May 1, 1990), www.doi.org/10.1084
     /jen.171.5.1705: Raziuddin, J.A. Mikovits, I Clavert, et al., "Negative Regulation of
     HIV-1 Expression in Monocytes: Role of the 65+50 kD NF kB Heterodimer," *Proceedings
     of the National Academy of Sciences* 88 (November 1, 1991): 9426-9430, www.doi
     .org/10.1073/pnas.88.21.9426, J.A. Mikovits, N. Lohrey, J. Cortless, et al., "Immune
     Activation of HIV Expression from Latently Infected Monocytes from Asymptomatic
     Seropositive Patients," *Journal of Clinical Investigation* 90 (October 1992); 1486-1491,
     www.doi.org/10.1172/JCI116016.

17   William R. Macon, Shyh-Ching Lo, Bernard J. Poiez, et al., "Acquired Immunodeficiency
     Syndrome-Like Illness with Systemic Mycoplasma Fermentans Infection in a Human
     Immunodeficiency Virus-Negative Man," *Human Pathology* 24 (May 1993): 554-558,
     www.doi.org/10.1016/0046-8177(93)90169-h.

18   Steven Epstein, *Impure Science: AIDS, Activism, and the Politics of Knowledge*, (Berkeley,
     CA University of California Press, 1998), www.drive.google.com/file/d/1AG1iFHZwCf
     GWClZzdni8GegBsWW6tISF/view

Chapter Eight

1   "Wernher Von Braun," *QuoteFancy* (Accessed July 16, 2021), www.quotefancy.com/
     quote/1104994/Wernher-von-Braun-Science-does-not-have-a-moral-dimension-It-is
     -like-a-knife-If-you-give

2   J.A. Mikovits, Raziuddin, M. Ruta, et al., "Negative Regulation of HIV Replication
     in Monocytes: Distinctions Between Restricted and Latent Expression in THP-1

Cells," *Journal of Experimental Medicine* 171 (May 1, 1990), www.doi.org/10.1084
/jen.171.5.1705, J.A. Mikovits, N. Lohrey, J. Cortless, et al., "Immune Activation of
HIV Expression from Latently Infected Monocytes from Asymptomatic Seropositive
Patients," *Journal of Clinical Investigation* 90 (October 1992); 1486-1491, www.doi.
org/10.1172/JCI116016.

3   John Crewdson, "Science Fictions: A Scientific Mystery, A Massive Coverup, and the
    Dark Legacy of Robert Gallo," Back Bay Books, New York, (2002), p. 539-540.

4   Paul W. Valentine, Researcher Sentenced in NIH Case," *Washington Post*, October 17,
    1992, www.washingtonpost.com/archive/politics/1992/10/17/researcher-sentenced-in
    -nih-case/e293929a-6062-445f-a503-7af1c862fd58/

5   "Biologist Accused Over Laboratory Funds," *New Scientist*, May 4, 1990, www
    .newscientist.com/article/mg12617151-400-biologist-accused-over-laboratory-funds/

6   E. Sitnicka, F. Ruscetti, G. Priestly, et al., "Transforming Growth Factor-B Directly and
    Reversibly Inhibits the Initial Cell Divisions of Long-Term Repopulating Hematopoietic
    Stem Calls," *Blood* 88 (Jly 1, 1996), 82-88.

7   J.A. Mikovits, H. Young, P. Vertino, et. al., "HIV-1 Infection Upregulates DNA
    Methyltransferase Resulting in De Novo Methylation of the IFN-y Promoter and
    Subsequent Downregulation of IFN-Production," *Molecular and Cellular* Biology 18
    (September 18, 1998), www.doi.org/10.1128/MCB.18.9.5166.

8   J.Y. Fang, J.A. Mikovits, R. Bagni, et al., "Infection of Lymphoid Cells by Replication
    -Defective HIV-1 Increases De Novo Methylation," *Journal of Virology* 75 (October
    2001): 9753-9761, www.doi.org/10.1128/JVI.75.20.9753-9761.2001.

Chapter Nine

1   "Carl Sagan Quotes," *GoodReads* (Accessed July 16, 2021), www.goodreads.com
    /quotes/142441-i-worry-that-especially-as-the-millennium-edges-nearer-pseudoscience

2   Syndney Brenner, "Retrospective: Frederick Sanger (1918-2013)," *Science* 17 (January
    17, 2014): 262, www.doi.org/10.1126/science.1249912.

3   K.S. Jones, C. Petrow-Sandowski, Y.K. Huang, et. Al., "Cell-Free HTLV-1 Infects
    Dendridic Cells Leading to Transmission and Transformation of CD4+T Cells," *Nature
    Medicine* 14 (April 14, 2008), 429-436, www.doi.org/10.1038/nm1745.

4   V.W. Valeri, A. Hryniewicz, V. Anderson, et. Al., "Requirement of the Human T-Cell
    Leukemia Virus p12 and p30 Products for Infectivity of Human Dendritic Cells and
    macaques, But Not Rabbits," *Blood* 116 (November 11, 2010) 3809-3817, www.doi.
    org/10.1182/blood-2010-05-284141.

5   C. Fenzia, M. Fiocchi, K. Jones, et. Al., "Human T-Cell Leukemia/Lymphoma Virus
    Type 1 p30, but not p12/p8, Counteracts Toll-Like Receptor 3 (TLR3) and TLR4
    Signaling in Human monocytes and Dendritic Cells," *Journal of Virology* 88 (January
    2014), www.doi.org/10.1128JVI.01788-13.

6   A. Bazarbach, Y. Plumelle, J.C. Ramos, et. al., "Meta Analysis on the Use of Zidovudine
    and Interferon Alfa in Adult T-Cell Leukemia/Lymphoma Showing Improved Survival
    in the Leukemic Subtypes," *Journal of Clinical Oncology* 28 (September 28, 20210)
    4177-4183, www.doi.org/10.1200?JCO.2010.28.0669.

7   R. Bagni, E. Barsov, B. Ortiz-Conde, et al., "Dendritic Cell-Meidated Infection of
    primary B Cells with KSHV," *Journal of Infectious Agents and Cancer* 4 (June 2009),
    www.doi.org/10.1186/1750-9378-4-S2-P9.

8   J.A. Kovacs, L. Deyton, R. Davey, et al., "Combined Zidovudine and Interferon-Alpha Therapy in Patients with Kaposi Sarcoma and the Acquired Immune Deficiency Syndrome (AIDS), *Annals of Internal Medicine* 111 (August 15, 1989), 280-287, www.doi.org/10.7326/0003-4819-111-4-280.

9   I. Busnadiego, S. Fernbach, M. Pohl, et al., "Antiviral Activity of Type I, II, and II Interferons Counterbalances ACE2 Inducibility and Restricts SARS0CoV-2," *MBIO* 11 (September-October 2020) e)1928-20, www.doi.org/10.1128/mBio.01928-20.

10 S. Bartelemez, C. Storey, P. Iversen, and F. Ruscetti, "Transient Inhibition of Endogenous Transforming Growth Factor 0 B1 (tgf-beta1) in Hematopoietic Stem Cells Accelerates Engraftment and Enhances Multi-Lineage Repopulating Efficiency," *Journal of Stem Cell Research and Therapeutics* I (December 20, 2016) 258-267, www.doi.org/10.15406/jsrt.2016.01.00045.

11 Ibid.

12 A.D. Bhatwadekar, E.P. Guerin, Y.P. Jarajapu, et al., ""Transient Inhibition of Transforming Growth Factor-Beta-1 in Human Diabetic CD34+ Cells Enhances Vascular Reparative Functions," *Diabetes* 50 (August 2010). www.doi.org/10.2337/db10-0287.

13 F.W. Ruscetti, D.A. Morgan, and R.C. Gallo, "Functional and Morphologic Characterization of Human T Cells Continuously Grown in Vitro," *Journal of Immunology* 119 (July 1977) 131-138.

14 V.C. Lombardi, F.W. Ruscetti, J. Das Gupta, et al., "Detection of an Infectious Retrovirus, XMRV, in Blood Cells of Patients with Chronic Fatigue Syndrome," *Science* 326 (October 23, 2009) 585-589, www.doi.org/10.1126/science.1179052.

15 Shyh-Ching Lo, Natalia Pripuzova, Binjie Li, et al., "Detection of MLV-Related Virus Gene Sequences in Blood of Patients with Chronic Fatigue Syndrome and Healthy Blood Donors," *Proceedings of the National Academy of Sciences* 107 (September 7, 2010) 15874-15879, www.doi.org/10.1073/pnas.1006901107.

16 "ANNOUNCEMENT: August 23, 2010 Human Gamma Retrovirus Test Now Available VIP Dx," http://www.vipdx.com/?forumid=331851.

17 Email from Allison Kanas to Judy Mikovits, December 2, 2011.

18 Email from Harold Varmus to Frank Ruscetti, September 26, 2010.

19 Shy-Ching Lo, Natalie Pripuzova, Bingjie Li, et al., "Retraction for Lo, et al., Detection of MLV-Related Virus Gene Sequences in Blood of Patients with Chronic Fatigue Syndrome and Healthy Blood Donors," *Proceedings of the National Academy of Sciences* 109 (January 3, 2012), 346, www.doi.org/10.1073/pnas.1119641109.

20 G.Q. Del Prete, M.F. Kearny, J. Spindler, et al., "Restricted Replication of Xenotropic Murine Leukemia Virus-Related Virus in Pigtailed Macaques," *Journal of Virology* (December 21, 2011) 3152-3166, www.doi.org/10.1128/JVI.06886-11.

21 Email from Maureen Hanson to Frank Ruscetti, September 10, 2013.

22 S. Panelli, L. Lorusso, A. Balestrieri, et al., "XMRV and Public Health: The Retroviral Genome is not a Suitable Template for Diagnostic PCR, and Its Association with Myalgic Encephalomyelitis/Chronic Fatigue Syndrome Appears Unreliable," *Frontiers in Public Health* 5 (May 22, 2017), www.doi.org/10.3389/fpubh.2017.00108.

23 John Coffin and Jonathon Stoye, "The Dangers of Xenotransplantation," *Nature*, (November 11, 1995, Letter to the Editor, vol.1, 1100.

24 Andrew Kolodny, "How FDA Failures Contributed to the Opioid Crisis," *AMA Journal of Ethics*, August 2020, www.journalofethics.ama-assn.org/article/how-fda-failures-contributed-opioid-crisis/2020-08

25 Jan Hoffman and Mary Williams Walsh, "Purdue Pharma Offers Plan to End Sackler Control and Mounting Lawsuits," *New York Times*, March 16, 2021, www.nytimes.com/2021/03/16/health/purdue-sacklers-bankruptcy-opioids.html

26 Kelly Brownell and Kenneth Warner, "The Perils of Ignoring History: Big Tobacco Played Dirty and Millions Died. How Similar is Big Food?" *Millibank Quaterly* 87 (March 2009), 259-294, www.doi.org/10.1111/j.1468-0009.2009.00555x.

27 "Women and HIV: A Spotlight on Adolescent Girls and Young Women," *UN AIDS*, (2019), www.unaids.org/sites/default/files/media_asset/2019_women-and-hiv_en.pdf

28 M. Rowbotham, W. Nothaft, R. Duan, et al., "Oral and Cutaneous Thermosensory profile of Selective TRPV1 Inhibition by ABT-102 in a Randomized Healthy Volunteer Trial," *Clinical Trial* 152 (May 2011) 1192-1200. www.doi.org/10.1016/j.pain.2011.01.051.

PART TWO

Prologue

1 CBS, Senior Member, "Suzanne Vernon: Agency Heads are Scared to Death . . . If XMRV Works Out," Phoenix Rising, February 23, 2011, www.forums.phoenixrising.me/threads/suzanne-vernon-agency-heads-are-scared-to-death-if-xmrv-works-out.8863/.

2 "Suzanne Vernon," *Encyclopedia of Myalgic Encephalomyelitis*, (Accessed June 28, 2021), www.me-pedia.org/wiki/Suzanne_Vernon.

3 Hillary Johnson, "Chasing the Shadow Virus," *Discover* Magazine, July 19, 2013, www.discovermagazine.com/health/chasing-the-shadow-virus-chronic-fatigue-syndrome-and-xmrv.

4 Shyh-Ching Lo, Natalia Pripuzova, Bingjie Li, et al, "Detection of MLV-Related Virus Gene Seqences of Blood in Patients with Chronic Fatigue Syndrome and Healthy Blood Donors," *Proceedings of the National Academy of Sciences*, August 23, 2010, doi:10.1073/pnas.100691107.

5 Ben Berkout, "Of Mice and Men: On the Origins of XMRV," *Frontiers in Microbiology*, Vol, 1, Article 147, (January 17, 2011), 4-5.

6 Email from Judy Mikovits to Simone Glynn and Frank Ruscetti, August 31, 2011, 8:24 PM, PDT.

7 Judy Mikovits Presentation at the New York Academy of Sciences, March 29, 2011, www.nyas.org/ebriefings/pathogens-in-the-blood-supply/.

8 Tara Haelle, "The INTERCEPT Blood System Rids Blood Donations of All Pathogens," July 1, 2015, *Scientific American*, www.scientificamerican.com/article/the-intercept-blood-system-rids-blood-donations-of-all-pathogens/

9 Jon Cohen, "The Waning Conflict Over XMRV and Chronic Fatigue Syndrome," *Science*, September 30, 2011, Vol. 333, Issue 6051, p. 1810, doi: 10.1126/science

10 Jeff German, "Harvey Whittemore Ordered to Surrender to Federal Prison Authorities," *Las Vegas Review*, June 5, 2014.

11 Jon Cohen, "Controversial CFS Researcher Arrested and Jailed," *Science* Magazine, November 19, 2011, www.sciencemag.org/news/2011/11/controversial-cfs-researcher-arrested-and-jailed.

12 Jon Cohen, "Embattled Institute Retains Major Grant to Study Chronic Fatigue Syndrome," *Science* Magazine, February 8, 2012, www.sciencemag.org/news/2012/02/embattled-institute-retains-major-grant-study-chronic-fatigue-syndrome.

[13] Martha Bellisle, "Wit and Work Made lobbyist Harvey Whittemore, '"An Institution,'" *Reno Gazette Journal*, February 12, 2012.

[14] Jon Cohen, "In a Rare Move, Science Without Authors' Consent Retracts paper that Tied Mouse Virus to Chronic Fatigue Syndrome," *Science* Magazine, December 22, 2011, www.sciencemag.org/news/2011/12/updated-rare-move-science-without -authors-consent-retracts-paper-tied-mouse-virus.

[15] Carl Zimmer, "A Man From Whom Viruses Can't Hide," November 22, 2010, *New York Times*, https://www.nytimes.com/2010/11/23/science/23prof.html.

[16] Ian Lipkin, Press Conference on Multi-Center Study at Columbia University, September 18, 2012.

[17] Ian Lipkin, Public Conference Call with the Centers for Disease Control, September 10, 2013. Transcript by ME/CFS Forums.com/wiki/Lipkin.

[18] Corvela Staff, "What Did We Find in the MMRV (Priorix Tetra) Vaccine?" Corvela, April 21, 2019, www.corvelva.it/en/speciale-corvelva/vaccinegate-en/what-did-we-find -in-the-mmrv-priorix-tetra-vaccine.html

[19] Meera Murgai, James Thomas, Olga Cherepanova, et al., "Xenotropic MLV Envelope Proteins Induce Tumor Cells to Secrete Factors that Promote the Formation of Immature Blood Vessels," *Retrovirology*, March 27, 2013, doi: 10.1186/1742-4690-10-34, www.pubmed.ncbi.nlm.nih.gov/23537062/.

[20] Meera Murgai, James Thomas, Olga Cherepanova, et al., "Xenotropic MLV Envelope Proteins Induce Tumor Cells to Secrete Factors that Promote the Formation of Immature Blood Vessels," *Retrovirology*, March 27, 2013, doi: 10.1186/1742-4690-10-34, www.pubmed.ncbi.nlm.nih.gov/23537062/.

Chapter One

[1] Letter from Dr. Joseph Gates for General Distribution, January 13, 1999.

[2] Email from Stephen B. Baylin, February 9, 2001.

Chapter Two

[1] "Interferon: The IF Drug for Cancer," *Time* magazine, March 31, 1980.

[2] Ibid.

[3] Ibid.

[4] Ibid.

[5] Val Hutchison and J.M. Cummins, "Low Dose Interferon in Patient with AIDS," The Lancet, Vol. 330, Issue 8574, pp 1530-1531 (December 26, 1987: doi: doi.org/10.1016 /S0140-6736(87)92671-7.

[6] Davy Koech, Arthur Obel, & Jun Minowada, et al., "Low-Dose Oral Alpha Interferon Therapy for Patients Seropositive for Human Immunodeficiency Virus Type-1 (HIV-1)," *Molecular Biotherapy*, Vol. 2, June 1990, pp. 91-95.

[7] Ibid.

[8] Gina Kolata, "Ignored AIDS Drug Shows Promise in Small Tests," *New York Times*, August 15, 1989.

[9] Ibid.

10   Emily Mantlo, Natalya Bukreyeva, Junki Maruyama, et al., "Antiviral Activities of Type I
     Interferons to SARS-CoV-2 Infection," *Journal of Antiviral Resistance*, Vol. 178, 104811
     (April 29, 2020): doi: 101016/j.antiviral.2020.104811.
11   Ivan Fan Hung, Kwok-Cheung Lung, et al., "Triple Combination of Interferon Beta 1b,
     Lopanavir-Ritonavir, and Ribavirin in the Treatment of Patients Admitted to hospital
     with COVID-19: An Open-Label, Randomized Phase 2 Trial," *The Lancet*, Vol. 395:
     1695-1704: doi: 1016/S0140-6736(20)31042-4.

Chapter Three

1   Phillip Hilts, "Federal Inquiry Finds Misconduct by a Discoverer of the AIDS Virus,"
     the *New York Times*, December 31, 1992.
2   Ibid.
3   "About Dr. Gallo," Institute of Human Virology, University of Maryland, www.ihv.org
     /about/About-Dr-Robert-C-Gallo/. (Accessed December 4, 2020).
4   Phillip Hilts, "Federal Inquiry Finds Misconduct by a Discoverer of the AIDS Virus,"
     the *New York Times*, December 31, 1992.
5   Ibid.
6   Douglas Kneeland, "The Gallo Case: A Three-Year Odyssey in Search of the Truth,"
     Chicago Tribune, December 6, 1992.
7   "Coronavirus Man-Made in Wuhan Lab, says Nobel Laureate," *The Week*, April 19,
     2020, www.theweek.in/news/world/2020/04/19/coronavirus-man-made-in-wuhan-lab
     -says-nobel-laureate.html.
8   Ibid.
9   John M. Coffin and Jonathon Stoye, "A New Virus for Old Diseases?" *Science*, Vol.
     326, issue 5952, pp 530-531, (October 23, 2009), doi: 10.1126/science.1181349,
     www.science.sciencemag.org/content/326/5952/530.full.
10   National Cancer Institute – HIV Drug Resistance Program – John M. Coffin, Ph.D.,
     Accessed September 13, 2013 www.home.ncifcr.gov/hivdrp/Coffin.html.
11   "Public Health Implications of XMRV Infection," Center for Cancer Research & Center
     of Excellence in HIV/AIDS and Cancer Virology – Workshop, July 22, 2009.
12   Ibid.
13   Paul Offit, "The Cutter Incident – How America's First Polio Vaccine Led to the Growing
     Vaccine Crisis," (New Haven and London; Yale University Press, 2005), 16.
14   "Specter of Paralysis Stalks Carolina," *The Literary Digest*, July 20, 1935.
15   Maurice Brodie, "Attempts to Produce Poliomyelitis in Refractory Lab Animals,"
     Experimental Biology and Medicine, (March 1, 1935), 832-836, doi: 10.3181/00379727
     -32-7876.
16   Ibid.
17   Maurice Brodie and William Park, "Active immunization Against Poliomyelitis,"
     American Journal of Public Health, (February 1936), 119-125.
18   G. Stuart, "The Problem of Mass Vaccination Against Yellow Fever," World Health
     Organization – Expert Committee on Yellow Fever, September 14-19, 1953, Kampala,
     Uganda.
19   Frank Ruscetti, Telephone Interview with Kent Heckenlively, June 14, 2012.

20   Jon Cohen and Martin Enserink, "False Positive," *Science*, Vol. 333, (September 23, 2011), 1694-1701.

21   Ibid.

22   Hillary Johnson, "Chasing the Shadow Virus: Chronic Fatigue Syndrome and XMRV," *Discover*, July 19, 2013, www.discovermagazine.com/health/chasing-the-shadow-virus-chronic-fatigue-syndrome-and-xmrv.

23   Zihao Yuan, Xuejun Fan, et al., "Presence of Complete Viral genome Sequences in Patient-Derived Xenografts," *Nature Communication*, May 1, 2021, doi: 10.1038/s41467-021-22200.5

24   Hillary Johnson, "Chasing the Shadow Virus: Chronic Fatigue Syndrome and XMRV," *Discover*, July 19, 2013, www.discovermagazine.com/health/chasing-the-shadow-virus-chronic-fatigue-syndrome-and-xmrv.

25   Delviks-Frankenberry K et al. *J. Virol.* 2013;87:11525-11537.

26   Ibid.

27   Ibid.

28   Paul Offit, "The Cutter Incident – How America's First Polio Vaccine Led to the Growing Vaccine Crisis," (New Haven and London; Yale University Press, 2005), 18.

29   Carl Zimmer, "A Man From Whom Viruses Can't Hide," the *New York Times*, November 22, 2010, www.nytimes.com/2010/11/23/science/23prof.html.

30   Ibid.

31   Hillary Johnson, "Chasing the Shadow Virus: Chronic Fatigue Syndrome and XMRV," *Discover*, July 19, 2013, www.discovermagazine.com/health/chasing-the-shadow-virus-chronic-fatigue-syndrome-and-xmrv.

32   Telephone Interview with Dr. Andrew Wakefield by Kent Heckenlively, February 25, 2016.

33   *In vivo* 25: 307-314 (2011).

34   Harvey Alter, Judy Mikovits, et al., "A Multicenter Blinded Analysis Indicates No Association Between Chronic Fatigue Syndrome/Myalgic Encephalomyelitis and Either Xenotropic Murine Leukemia Virus0Related Virus and Polytropic Murine Leukemia Virus," MBIO, September 18, 2012, doi: 10.1128/mBio.00266-12, www.ncbi.nlm.nih.gov/pmc/articles/PMC3448165/.

35   Telephone Interview with Paul Cheney by Kent Heckenlively, July 25, 2013.

36   Ibid.

37   Ibid.

38   "CII's W. Ian Lipkin Receives NIH Grant to Establish a New Center," Columbia University, March 10, 2014, www.publichealth.columbia.edu/research/center-infection-and-immunity/ciis-w-ian-lipkin-receives-nih-grant-establish-new-center.

39   Amy Qin and Chris Buckley, "A Top Virologist, at Center of a Pandemic Storm, Speaks Out," *New York Times*, June 14, 2021, www.nytimes.com/2021/06/14/world/asia/china-covid-wuhan-lab-leak.html.

40   Ibid.

41   Ibid.

42   Ibid.

43   Harvey Alter, Judy Mikovits, et al., "A Multicenter Blinded Analysis Indicates No Association Between Chronic Fatigue Syndrome/Myalgic Encephalomyelitis and Either Xenotropic Murine Leukemia Virus0Related Virus and Polytropic Murine Leukemia Virus," MBIO, September 18, 2012, doi: 10.1128/mBio.00266-12, www.ncbi.nlm.nih.gov/pmc/articles/PMC3448165/.

44 Meredith Wadman, "Lawsuit at Columbia University Roils Prominent Chronic Fatigue Syndrome Research Lab," *Science*, May 23, 2017, www.sciencemag.org /news/2017/05/lawsuit-columbia-university-roils-prominent-chronic-fatigue -syndrome-research-lab.

45 Ibid.

46 Ibid.

47 Joe Schoffstall, "Columbia Professor Who Thanked Fauci for Wuhan Lab Messaging has Links to Chinese Communist Party Members," Fox News, July 1, 2021, www.foxnews. com/politics/columbia-professor-lipkin-fauci-wuhan-lab-china.

48 "About Harold," Laboratory of Harold Varmus, www.varmuslab.org/about-harold-2, (Accessed December 9, 2020).

49 "Budget," National Institutes of Health, www.nih.gov/about-nih/what-we-do/budget, (Accessed December 9, 2020).

50 "List of NIH Institutes, Centers, and Offices," www.nih.gov/institutes-nih/list-nih -institutes-centers-offices, (Accessed December 9, 2020).

51 "Sloan Kettering CEO Craig Thompson's $6.7 Million Pay," Future of Capitalism, October 3, 2018, www.futureofcapitalism.com/2018/10/sloan-kettering-ceo-craig -thompson-67-million-pay.

52 "National Institutes of Health," Federal Pay, www.federalpay.org/employees/national -institutes-of-health/top-100/2017, (Accessed December 9, 2009).

53 "W. Ian Lipkin," *The Lancet COVID-19 Commission*, www.covid19commission.org/ian -lipkin (Accessed December 13, 2020).

Chapter Four

1 *J Clin Invest.* 2020;130(5):2347-2363. https://doi.org/10.1172/JCI122462.

2 Ibid.

3 Telephone Interview of Michael Lourdes by Kent Heckenlively, December 4, 2020.

4 Ibid.

5 Ibid.

6 Ibid.

7 Ibid.

8 Ibid.

9 Ibid.

10 Ibid.

11 Ibid.

12 Ibid.

13 Ian Lipkin, Public Conference Call with the Centers for Disease Control, September 10, 2013. Transcript by ME/CFS Forums.com/wiki/lipkin.

14 Telephone Interview of Michael Lourdes by Kent Heckenlively, December 4, 2020.

15 Ibid.

Chapter Five

1 L. Montagnier & F.K. Sanders, "Replicative Form of Encephalomyocarditis Virus Ribonucleic Acid," *Nature*, Vol. 199, pp. 664-667, (August 17, 1963), doi: 10.1038/199664a0, www.pubmed.ncbi.nlm.nih.gov/14074552/.

2   Luc Montagnier, "Nobel Prize Biographical," Nobel Prize Committee, Delivered December
    7, 2008, www.nobelprize.org/prizes/medicine/2008/montagnier/biographical/.
3   Telephone Interview with Luc Montagnier by Kent Heckenlively, January 17, 2021.
4   Luc Montagnier, "Virus," (1990, English translation by W. W. Norton, 2000), p.81
5   Telephone Interview with Luc Montagnier by Kent Heckenlively, January 17, 2021.
6   Luc Montagnier, "Virus," (1990, English translation by W. W. Norton, 2000), p.119.
7   Ibid at p. 121.
8   "French Nobel Prize Winner: COVID-19 made in a Lab," *Connexion France*, April 22,
    2020,   www.connexionfrance.com/French-news/Disputed-French-Nobel-winner-Luc
    -Montagnier-says-Covid-19-was-made-in-a-lab-laboratory.
9   Luc Montagnier, "Nobel Prize Biographical," Nobel Prize Committee, December 7,
    2008, www.nobelprize.org/prizes/medicine/2008/montagnier/biographical/.
10 Telephone Interview with Luc Montagnier by Kent Heckenlively, January 17, 2021.
11 Ibid.
12 Ibid.

Chapter Six

1   "Stephanie Seneff, "Curriculum Vitae," September 22, 2020.
2   Telephone Interview with Stephanie Seneff by Kent Heckenlively, January 11, 2021.
3   Ibid.
4   Ibid.
5   Ibid.
6   Ibid.
7   Stephanie Seneff and Laura Orlando, "Glyphosate Substitution for Glycine During Protein
    Synthesis as a Causal Factor in Mesoamerican Neuropathy," Journal of Environmental &
    Analytical Toxicology, Vol. 8, Issue 1: doi:10.4172/2161-0525.1000541.
8   Telephone Interview with Stephanie Seneff by Kent Heckenlively, January 11, 2021.
9   Ibid.
10 Steven A. Kemp, Dami A. Collier, Rawlings P. Datir, et al., "SARS-CoV-2 Evolution
    During Treatment of Chronic Infection," *Nature*, (Case Reports), Vol. 592(7853);
    p. 277-282: doi: 10.1038/s41586/s41586-021-03291-y.
11 Telephone Interview with Stephanie Seneff by Kent Heckenlively, January 11, 2021.
12 Chien-Te Tseng, Elena Sbrana, Naoko Iwata-Yoshikawa, et al., "Immunization with
    SARS Coronavirus Vaccines Leads to Pulmonary Immunopathology on Challenge
    with the SARS Virus," *PLOSOne*, April 20, 2012: doi.org/10.1371/journal.
    pone.0035421.
13 Telephone Interview with Stephanie Seneff by Kent Heckenlively, January 11, 2021.
14 Jason B. Wilson, Iraj Khabazian, Margaret Wong, et al., "Behavioral and Neurological
    Correlates of ALS-Parkinsonism Dementia Complex in Adult Mice Fed Washed Cycad
    Flour," *NeuroMolecular Medicine*, Vol. 1, pp. 207-221, (2002).
15 Michael Petrick, Margaret Wong, Rena Tabata, et. Al., "Aluminum Adjuvant Linked to
    Gulf War Illness Induces Motor Neuron Death in Mice," Neuromolecular Medicine,
    Vol. 9(1), pp. 83-100, (2007), doi: 10.1385/nmm:9:1:83.
16 Telephone Interview with Dr. Christopher Shaw by Kent Heckenlively, January 8, 2021.
17 Ibid.
18 Ibid.

[19] Housam Eidi, Janice Yoo, Suresh Bairwa, et. al. "Early Postnatal Injections of Whole Vaccines Compared to Placebo Controls: Differential Behavioral Outcomes in Mice," Journal of Inorganic Biochemistry, Vol. 212, (November 2020); doi: 10.1016/jinorgbio.2020.11120.

[20] Telephone Interview with Dr. Christopher Shaw by Kent Heckenlively, January 8, 2021.

[21] Ibid.

[22] Ibid.

[23] Ibid.

[24] Telephone Interview with Dr. Frank Shallenberger by Kent Heckenlively, October 30, 2020.

[25] Ibid.

[26] Ibid.

[27] Ibid.

[28] Ibid.

[29] Stephen Levine and Parris Kidd, "Antioxidant Adaptation," Biocurrents Publishing, (January 1, 1986).

[30] Telephone Interview with Dr. Frank Shallenberger by Kent Heckenlively, October 30, 2020.

[31] Ibid.

[32] Frank Shallenberger, "The Ozone Miracle: How You Can Harness the Power of Oxygen to Keep You and Your Family Healthy," Create Space Independent Publishing Platform, p. 11, March 27, 2017.

[33] M.T. Carpendale & J.K. Freeberg, "Ozone Inactivates HIV at Non-Cytotoxic Concentrations," Journal of Antiviral Resistance, October 1991, vol. 16(3):281-92, doi: 10.1016/0166-3542(91)90007-e.

[34] Telephone Interview with Dr. Frank Shallenberger by Kent Heckenlively, October 30, 2020.

[35] Ibid.

[36] Frank Shallenberger, "Selective Compartmental Dominance: An Explanation for a noninfectious, Multifactorial Etiology for Acquired Immune Deficiency Syndrome (AIDS), and a Rationale for Ozone Therapy and Other Immune Modulating Therapies," Medical Hypotheses, (1998) Vol. 50, p. 67-80.

[37] M. Clerici & G.M. Shearer, "A TH1 to TH@ Switch is a Critical Step in the Etiology of HIV Infection," Immunology Today, March 14, 1993 vol. 3: 107-111, doi: 10.1016/0167-5699(93)90208-3.

[38] Telephone Interview with Dr. Frank Shallenberger by Kent Heckenlively, October 30, 2020.

[39] Ibid.

[40] Ibid.

[41] Ibid.

[42] Ibid.

[43] Ibid.

Epilogue

[1] Judy Mikovits, "The Exotic Biology of Xenotropic Murine Leukemia Viruses (XMRVs)," Plague the Book website, November 2013, www.plaguethebook.com/wp-content /uploads/2019/06/The-Exotic-Biology-of-XMRV.pdf.

2   Anatoly Urisman, Ross J. Molinaro, et al., "Identification of a Novel Gammaretrovirus in Prostate Tumors of Patients Homozygous for R462Q RNASEL Variant," *PLOS Pathogen*, March 31, 2006, doi: 101371/journal.ppat0020025, www.pubmed.ncbi.nlm .nih.gov/16609730/.

3   Yu-Ann Zhang, Anirban Maitra, et al., "Frequent Detection of Infectious Xenotropic Murine Leukemia Virus (XMLV) in Human Cultures Established from Mouse Xenografts," *Journal of Cancer Biology and Therapeutics*, October 11, 2011, vol. 12(7), p. 617-628, doi: 10.4161/cbt.12.715955, www.pubmed.ncbi.nlm.nih.gov/21750403/.

4   Meera Murgai, James Thomas, et al., "Xenotropic MLV Envelope Proteins Induce Tumor Cells to Secrete Factors that Promote the Formation of Immature Blood Cells," *Journal of Retrovirology*, doi: 10.1186/1742-4690-10-34 October 1, 2013, www.retrovirology .biomedcentral.com/articles/10.1186/1742-4690-10-34.

5   Judy Mikovits, "The Exotic Biology of Xenotropic Murine Leukemia Viruses (XMRVs)," *Plague the Book* website, November 2013, www.plaguethebook.com/wp-content /uploads/2019/06/The-Exotic-Biology-of-XMRV.pdf.

6   Krista Delvkis-Frankenberry, Tobias Paprotka, et al., "Generation of Multiple Replication-Competent Retroviruses through Recombination between PreXMRV-1 and PreXMRV-2," *Journal of Virology*, Vol. 87(21), p. 11525-11537, November 2013, doi: 10.1128/JVI.01787-13, www.ncbi.nlm.nih.gov/pmc/articles/PMC3807343/

7   Judy Mikovits, "The Exotic Biology of Xenotropic Murine Leukemia Viruses (XMRVs)," *Plague the Book* website, November 2013, www.plaguethebook.com/wp-content /uploads/2019/06/The-Exotic-Biology-of-XMRV.pdf.

8   Ibid.

9   Esther Nolte-'t Hoen, Tom Cremer, et al., "Extracellular Vesicles and Viruses: Are They Close Relatives?" *Proceedings of the National Academy of Sciences – Perspective*, Vol. 113, No. 33, p. 9155-9161, August 16, 2016, doi: 10.1073/pnas.1605146113, www.ncbi .nlm.nih.gov/pmc/articles/PMC4995926/.

10 Ibid at 9155.

11 Stuart Neil and Edward Campbell, "Fake Science: XMRV: COVID-19, and the Toxic legacy of Dr. Judy Mikovits," *AIDS Research and Human Retroviruses*, Vol. 36, No. 7, p. 45-549, May 22, 2020, doi: 10.1089/AID.2020.0095. www.pubmed.ncbi.nlm.nih .gov/32414291/.

12 Declan Butler, "Engineered Bat Virus Stirs Debate Over Risky Research," *Nature*, November 12, 2015, doi: 10.1038/Nature.2015.18787.

13 Ibid.

14 Brendan Morrow, "Former CDC Director Surprises CNN's Sanjay Gupta by Revealing He Believes COVID-19 Originated in a Wuhan Lab," *CNN*, March 26, 2021, www .news.yahoo.com/former-cdc-director-surprises-cnns-151700805.html.

15 "Buzz Photos, Freedom Watch and Larry Klayman vs. The People's Republic of China, the People's Liberation Army, and the Wuhan Institute of Virology," United States District Court for the Northern District of Texas, March 17, 2020, www.freedomwatchusa.org /pdf/200317-CoronavirusFILEDComplaint177113137478.pdf.

Acknowledgments

From Frank:

Scientists like to pretend, especially when awards are doled out, that they did it all on their own. At least in my case, nothing could be further from the truth. From the very beginning I was a moon borrowing light from many suns. In the service, Chuck Coltman and Dorothy Grisham. In graduate school, Lew and Linda Jacobson. All the wonderful students, Carol Henry, Sue Leschine, Kathy Christ, and Roger Weppelman. In the medical school, Dane and Sally Boggs, Beverley Torok, Joan Turner, and Bernie Fisher. At NCI, Andrea Woods, Linda and Steve Hunt, Ray Kiefer, Corrado Tarella, Frederick, Joost Oppenheim, Dan Longo, Craig Reynolds, Howard Young, Linda and Dave Wolff, Nancy Colburn, Cho-Chi Li, Cari Sadowski, Dan Bertolette, and Kathy Jones. Cheers again to all the wonderful people I mentioned in the text. Special praise to my modern dream chasers, Steve Bartelmez, Pat Iverson, and Charlie Garcia. To Dr. J without whom this book was impossible.

And, as always, I have to thank my wife, Sandra Kay, who, the more I listen to her the more I draw inspiration from her humanity, wisdom, and compassion, and to my son who is taking after his mother in all the right ways.

From Judy:

I would like to thank my family for their unwavering support throughout this very difficult decade. My parents Gloria Furr-Fornshill and John L Mikovits Jr. and my siblings John, Julie, and Karen and their families fought for me with every resource they had and more importantly never blamed me for the humiliation and dishonor this plague of corruption brought to our family name and our proud Cherokee and Austrian/Hungarian heritage. My mom raised us alone from the age of ten and taught us above all else integrity and honesty. To mom the most egregious offense was to be silent about any wrong. To witness a wrong and do nothing generated explosive anger and guaranteed fierce punishment. As she lay dying earlier this year, I could get a laugh by reminding her the worst of punishments were usually precipitated by the statement "God gave you a mouth, USE IT" and in later years her beloved husband Ken would say to any who would listen "Judy's mouth gets Judy's body in trouble"; to which mom would beam with pride. I likely would not have survived without my dear husband David. There simply is no kinder human being. Few men would willingly lose everything to keep their wife out of jail. Whenever I get discouraged he texts me two songs: Marilyn McBride's "Anyway" and Train's "Calling all Angels."

I would also like to thank my church families at Community Presbyterian Church in Ventura, CA and North Coast Church in Carlsbad, CA and my friends at Pierpont Bay Yacht Club. In particular, the Stephens Ministry Program at CPC and North Coast Pastor Larry Osborne's book *Thriving in Babylon*. Without the love and teaching of these dear friends who never left my side during the darkest of times, I might never have endured. They were often the only light in this very dark decade.

Special thanks to Lois Hart and Robyn E (My Cherokee Twin) and Travis Middleton who worked tirelessly without payment for now almost a decade trying to bring me justice in the legal system in Nevada detailing the crimes and obstruction of justice, which allowed the government to perpetrate these crimes against humanity.

I grieve the loss of colleagues and the many new friends who suffer from these devastating diseases and the deaths of their loved ones from ME/CFS, autism, and cancer. I know that no words can bring back the lost decades and loved ones but hope that this book can end the stigma and help heal their families from wounds too deep to imagine. Each and every one are my heroes and heroines. I cannot name them for fear they will face additional retribution.

From Kent:

I'd first like to thank my wonderful partner in life, Linda, and our two children Jacqueline and Ben, for their constant love and support. I'd like to thank my mother, Josephine, and my father, Jack, for teaching me to tell the truth regardless of the consequences and showing how to love through even difficult times. I'd like to thank the best brother in the world, Jay, and his wonderful wife, Andrea, and their three children, Anna, John, and Laura, for always being on my side.

I'd like to thank some of the wonderful teachers in my life, my seventh- grade science teacher, Paul Rago, my eighth-grade English teacher, Elizabeth White, my high school science teacher, Ed Balsdon, my religion teacher Brother Richard Orona, and in college, English professors Clinton Bond, Robert Haas, Carol Lashoff, and in the political science department, David Alvarez, who nominated me to be the school's Rhodes Scholar candidate. I'd also like to thank my college rowing coach, Giancarlo Trevisan, the mad Italian, who showed me what it means to have crazy passion for an often-overlooked sport. In law school, I'd like to thank Bernie Segal, the criminal defense attorney who taught me to always have hope that justice will eventually prevail. I'd like to thank my writing teachers, James Frey, who looked at me one time and said, "Yeah, I think you'll be a writer," as well as Donna Levin, and James Dalessandro, who always said to find the story first, then write the hell out of it.

My life wouldn't be complete without my great friends, John Wible, John Henry, Pete Klenow, Chris Sweeney, Suzanne Golibart, Gina Cioffi, Eric Holm, Susanne Brown, Rick Friedling, Max Swafford, Sherilyn Todd, Rick and Robin Kreutzer, Christie and Joaquim Pereira, Tricia Mangiapane, and all of you who have made my passage through life such a party.

I work with the best group of science teachers at Gale Ranch Middle School, Danielle Pisa, Neelam Bhokani, Amelia Larson, Matt Lundberg, Katie Strube, Derek Augarten, and Arash Pakhdal. Thanks for always challenging my thinking and making me ask what is best for our students.

In the activist community, I'd like to thank J.B. Handley, Tim Bolen, Mary Holland, Lou Conte, Del Bigtree, Brian Hooker, Barry Segal, Elizabeth Horn, Brian Burrowes, Polly Tommey, Dr. Andrew Wakefield, and Robert F. Kennedy, Jr., for their friendship in the continuing fight against the Goliath of corrupted science.

Lastly, I'd like to thank my agent, Johanna Maaghul, my wonderful editors, Anna Wostenberg, and at Skyhorse the fabulous Caroline Russomanno, and for the faith shown in me over the years by publisher Tony Lyons.